ENVIRONMENTAL SCIENCE, ENGINEERING
AND TECHNOLOGY SERIES

MECHANISMS OF CADMIUM TOXICITY TO VARIOUS TROPHIC SALTWATER ORGANISMS

ENVIRONMENTAL SCIENCE, ENGINEERING AND TECHNOLOGY SERIES

Nitrous Oxide Emissions Research Progress
Adam I. Sheldon and Edward P. Barnhart (Editors)
2009. ISBN: 978-1-60692-267-5

Fundamentals and Applications of Biosorption Isotherms, Kinetics and Thermodynamics
Yu Liu and Jianlong Wang (Editors)
2009. ISBN: 978-1-60741-169-7

Environmental Effects of Off-Highway Vehicles
Douglas S. Ouren, Christopher Haas, Cynthia P. Melcher, Susan C. Stewart, Phadrea D. Ponds, Natalie R. Sexton, Lucy Burris, Tammy Fancher and Zachary H. Bowen
2009. ISBN: 978-1-60692-936-0

Agricultural Runoff, Coastal Engineering and Flooding
Christopher A. Hudspeth and Timothy E. Reeve (Editors)
2009. ISBN: 978-1-60741-097-3

Agricultural Runoff, Coastal Engineering and Flooding
Christopher A. Hudspeth and Timothy E. Reeve (Editors)
2009. ISBN: 978-1-60876-608-6

Conservation of Natural Resources
Nikolas J. Kudrow (Editor)
2009. ISBN: 978-1-60741-178-9

Conservation of Natural Resources
Nikolas J. Kudrow (Editor)
2009. ISBN: 978-1-60876-642-6
(Online Book)

Directory of Conservation Funding Sources for Developing Countries: Conservation Biology, Education and Training, Fellowships and Scholarships
Alfred O. Owino and Joseph O. Oyugi
2009. ISBN: 978-1-60741-367-7

Forest Canopies: Forest Production, Ecosystem Health and Climate Conditions
Jason D. Creighton and Paul J. Roney (Editors)
2009. ISBN: 978-1-60741-457-5

Soil Fertility
Derek P. Lucero and Joseph E. Boggs (Editors)
2009. ISBN: 978-1-60741-466-7

The Amazon Gold Rush and Environmental Mercury Contamination
Daniel Marcos Bonotto and Ene Glória da Silveira
2009. ISBN: 978-1-60741-609-8

Buildings and the Environment
Jonas Nemecek and Patrik Schulz (Editors)
2009. ISBN: 978-1-60876-128-9

Tree Growth: Influences, Layers and Types
Wesley P. Karam (Editor)
2009. ISBN: 978-1-60741-784-2

Syngas: Production Methods, Post Treatment and Economics
Adorjan Kurucz and Izsak Bencik (Editors)
2009. ISBN: 978-1-60741-841-2

Syngas: Production Methods, Post Treatment and Economics
Adorjan Kurucz and Izsak Bencik (Editors)
2009. ISBN: 978-1-61668-214-9
(Online Book)

Process Engineering in Plant-Based Products
Hongzhang Chen
2009. ISBN: 978-1-60741-962-4

Carbon Capture and Storage including Coal-Fired Power Plants
Todd P. Carington (Editor)
2009. ISBN: 978-1-60741-196-3

Potential of Activated Sludge Utilization
Xiaoyi Yang
2010. ISBN: 978-1-60876-019-0

Psychological Approaches to Sustainability: Current Trends in Theory, Research and Applications
Victor Corral-Verdugo, Cirilo H. Garcia-Cadena and Martha Frias-Armenta (Editors)
2010. ISBN: 978-1-60876-356-6

Fluid Waste Disposal
Kay W. Canton (Editor)
2010. ISBN: 978-1-60741-915-0

Recent Progress on Earthquake Geology
Pierpaolo Guarnieri (Editor)
2010. ISBN: 978-1-60876-147-0

Check Dams, Morphological Adjustments and Erosion Control in Torrential Streams
Carmelo Consesa Garcia and Mario Aristide Lenzi (Editors)
2010. ISBN: 978-1-60876-146-3

**Freshwater Ecosystems
and Aquaculture Research**
*Felice De Carlo and Alessio Bassano
(Editors)*
2010. ISBN: 978-1-60741-707-1

**Grassland Biodiversity: Habitat
Types, Ecological Processes
and Environmental Impacts**
*Johan Runas and Theodor Dahlgren
(Editors)*
2010. ISBN: 978-1-60876-542-3

Handbook of Environmental Policy
*Johannes Meijer and Arjan der Berg
(Editors)*
2010. ISBN: 978-1-60741-635-7

Environmental Modeling with GPS
Lubos Matejicek (Editor)
2010. ISBN: 978-1-60876-363-4

**Handbook on Agroforestry:
Management Practices
and Environmental Impact**
Lawrence R. Kellimore (Editor)
2010. ISBN: 978-1-60876-359-7

**Pipelines for Carbon
Sequestration: Background
and Issues**
Elvira S. Hoffmann (Editor)
2010. ISBN: 978-1-60741-383-7

Biodiversity Hotspots
*Vittore Rescigno and Savario Maletta
(Editors)*
2010. ISBN: 978-1-60876-458-7

**Mechanisms of Cadmium
Toxicity to Various
Trophic Saltwater Organisms**
*Zaosheng Wang, Changzhou Yan,
Hainan Kong and Deyi Wu*
2010. ISBN: 978-1-60876-646-8

ENVIRONMENTAL SCIENCE, ENGINEERING
AND TECHNOLOGY SERIES

MECHANISMS OF CADMIUM TOXICITY TO VARIOUS TROPHIC SALTWATER ORGANISMS

ZAOSHENG WANG,
CHANGZHOU YAN,
HAINAN KONG
AND
DEYI WU

Nova Science Publishers, Inc.
New York

Copyright © 2010 by Nova Science Publishers, Inc.

All rights reserved. No part of this book may be reproduced, stored in a retrieval system or transmitted in any form or by any means: electronic, electrostatic, magnetic, tape, mechanical photocopying, recording or otherwise without the written permission of the Publisher.

For permission to use material from this book please contact us:
Telephone 631-231-7269; Fax 631-231-8175
Web Site: http://www.novapublishers.com

NOTICE TO THE READER
The Publisher has taken reasonable care in the preparation of this book, but makes no expressed or implied warranty of any kind and assumes no responsibility for any errors or omissions. No liability is assumed for incidental or consequential damages in connection with or arising out of information contained in this book. The Publisher shall not be liable for any special, consequential, or exemplary damages resulting, in whole or in part, from the readers' use of, or reliance upon, this material. Any parts of this book based on government reports are so indicated and copyright is claimed for those parts to the extent applicable to compilations of such works.

Independent verification should be sought for any data, advice or recommendations contained in this book. In addition, no responsibility is assumed by the publisher for any injury and/or damage to persons or property arising from any methods, products, instructions, ideas or otherwise contained in this publication.

This publication is designed to provide accurate and authoritative information with regard to the subject matter covered herein. It is sold with the clear understanding that the Publisher is not engaged in rendering legal or any other professional services. If legal or any other expert assistance is required, the services of a competent person should be sought. FROM A DECLARATION OF PARTICIPANTS JOINTLY ADOPTED BY A COMMITTEE OF THE AMERICAN BAR ASSOCIATION AND A COMMITTEE OF PUBLISHERS.

LIBRARY OF CONGRESS CATALOGING-IN-PUBLICATION DATA
Mechanisms of cadmium toxicity to various trophic saltwater organisms / Zaosheng Wang ... [et al.].
   p. cm.
  Includes index.
  ISBN 978-1-60876-646-8 (softcover)
  1. Cadmium--Toxicology. 2. Cadmium--Environmental aspects. 3. Aquatic organisms--Effect of water pollution on. I. Wang, Zaosheng, 1965-
  RA1231.C3M43 2009
  615.9'25662--dc22   2009041966

Published by Nova Science Publishers, Inc. ✛ New York

Contents

Preface		ix
Chapter 1	Mechanisms of Cadmium Toxicity to Various Trophic Saltwater Organisms	1
Chapter 2	Sources and Pathways of Cadmium in the Environment	3
Chapter 3	Toxicity of Waterborne Cadmium to Saltwater Aquatic Organisms	15
Chapter 4	Toxicity of Dietary Cadmium to Aquatic Organisms	33
Chapter 5	Environmental Safety of Cadmium	47
Chapter 6	Conclusion	51
Chapter 7	Perspectives	53
Acknowledgments		55
Abbreviations		57
References		59
Index		65

PREFACE

In this chapter, we summarize basic and experimental knowledge on sources and toxicity of cadmium and overview mechanisms of toxicity and detoxification for various trophic aquatic organisms exposed to cadmium different routes, which will assist ecological risk assessments involving cadmium toxicity. Cadmium is known to be both extremely toxic and ubiquitous in natural environments. It exists most commonly as a trace constituent in natural ecosystems, where its natural occurrence appears to not cause harm to the environment. However, it can be mobilized by a number of processes especially human activities. Anthropogenic sources of cadmium have contributed large amounts of this potentially toxic metal to the water cycle with ultimate impacts on aquatic ecosystems.When toxicity values are compared among different phyla or classes of various trophic aquatic organisms including Annelida (polychaete worm, oligochaete worm), Mollusca (oyster, mussel, clam, snail, squid), Arthropoda (copepod, cladoceran, amphipod, mysid, crab, shrimp, lobster, iosopd), Echinodermata (sea urchin), and Chordata (osteichthyes, fish), species sensitivity distributions (SSDs) can be obtained. Concentrations of cadmium that induce toxicity vary over several orders of magnitude among invertebrate species, with the mysid shrimp (Crustacea) reported as the most sensitive. Furthermore, an example of chronic toxic effects is given with the saltwater cladoceran *Moina monogolica* Daday exposed to dissolved and dietary routes of cadmium. The mechanisms of cadmium accumulation patterns including uptake and elimination have been analyzed, with the underlying toxicity, accumulation and critical body residues (CBRs) of cadmium being species-specific, with further interspecies differences occurring in cadmium uptake and elimination rates, in which residues of cadmium in organisms or tissues are linked to organismal responses, cadmium detoxification involves metal rich granules (MRG) playing a role in tolerance to long-term cadmium exposure, while metallothiooneins (MTs) mainly act against short-term cadmium exposure. This may cause a distinction between mechanisms of acute and chronic cadmium toxicity. Chronic toxicity could result from a "spill over" effect and the overwhelming of induced detoxification mechanisms over a period of time. In response to the deleterious effects of cadmium present in the environment compared with threshold values, it is important to establish more strict norms to protect most of the species against cadmium toxicity (e.g. 95%). Characterizing the estuarine and marine environments in which organisms are exposed could improve predictions of cadmium fate and transport and ultimately toxicity and mechanisms of cadmium to aquatic biota, which is especially important in populated, urban environments considering the increasing urban development and extensive utilization of cadmium.

Chapter 1

MECHANISMS OF CADMIUM TOXICITY TO VARIOUS TROPHIC SALTWATER ORGANISMS

1. INTRODUCTION

Metal pollution of the marine environment is a major problem of increasing magnitude that has become an issue of concern, because most of the metals are transported into the marine environment and accumulated without decomposition. GESAMP (the joint Group of Experts on the Scientific Aspects of Marine Pollution) defined marine pollution as "introduction by man directly or indirectly, of substances or energy into the marine environment (including estuaries) resulting in such deleterious effects as harm to living resources, hazards to human health, hindrance to marine activities including fishing, impairment of quality for use of sea-water, and reduction of amenities."

Cadmium is one of the metals whose concentration is increasing in aquatic environments, and characteristic of Group IIB elements (of the Periodic Table (Zn, Cd, Hg)) is scarce, fairly expensive and of low mechanical strength. It has atomic number 48 and an atomic weight of 112.40. Cadmium was first isolated and identified from the zinc ore smithsonite ($ZnCO_3$), and over centuries has been released slowly into the environment from widespread sources such as the smelting of a variety of ores and the burning of wood and coal.

Cadmium can enter the environment from various anthropogenic sources such as by-products from zinc refining, coal combustion, mine wastes, electroplating processes, iron and steel production, pigments, fertilizers and pesticides. Through quantifying the relative importance of anthropogenic and natural contributions to cadmium cycling in the environment for both pre- and post-industrial times, it has been consistently shown that the contribution from anthropogenic sources (e.g. non-ferrous metal industry) has increased greatly over the past century and currently dominates the cadmium biogeochemical cycle (Campbell, 2006).

Cadmium has been ranked as one of the major metal hazards, because there is now mounting evidence that it is present in aquatic and terrestrial environments at levels that are sufficient to produce biological effects to various organisms. Cadmium exerts harmful effects on aquatic organisms in many ways, although all the major mechanisms of toxicity are a consequence of the strong coordinating properties of cadmium cations [Cd^{2+}] that affect the properties of many biological molecules (enzymes, etc.), often by blocking and reducing the thiol sites on proteins (Kneer and Zenk, 1992). Moreover, cadmium can be accumulated via the food chain, posing a serious threat to human health (Vijver et al., 2005).

Prediction of metal bioaccumulation and toxicity in aquatic organisms has been based on the free ion activity (e.g., free ion activity model; Campbell, 1995) or more recently on the binding with the biological/toxicological sites of action (e.g., biotic ligand model; Paquin et al., 2002). However, metals are bound to various intracellular ligands that may control metal toxicity. Among the five operationally defined subcellular fractions, namely metal-rich granules, cellular debris, organelles, heat-denatured protein (HDP), and heat-stable protein (HSP), cadmium was mostly bound to HSP, whereas it was least bound to HDP. Cadmium was redistributed with increasing [Cd^{2+}] concentration from the biologically detoxified pool to the presumed metal-sensitive fractions (MSF, a combination of organelles and HDP), which led to higher cellular cadmium accumulation, toxicity, and sensitivity. The MSF can provide the better predictor of cadmium toxicity than [Cd^{2+}] concentration or cellular accumulation, indicating that models predicting cadmium toxicity need to address the subcellular fate of cadmium and how this responds to external and internal conditions (Wang and Wang, 2008; Miao and Wang, 2006).

In this chapter, the bioaccumulation, sub-cellular distribution, and toxicity of cadmium in aquatic organisms is described.

Overall, this chapter underlines the need to better understand the uptake sites for dietary and/or waterborne exposure, and at how cadmium is taken up, eliminated and detoxified. This will improve our ability to predict the potential for toxic effects from aquatic organism exposure to environmentally realistic cadmium concentrations.

Chapter 2

SOURCES AND PATHWAYS OF CADMIUM IN THE ENVIRONMENT

As already noted, cadmium may be present in the environment both from natural and anthropogenic sources. It occurs mainly as a component of minerals in the earth's crust at an average concentration of 0.18 mg/kg and usually ranges from approximately 0.01 to 1.8 mg/kg in soils. In natural freshwaters, cadmium can occur at concentrations below 0.1 µg Cd/L, but in environments impacted by human activities, concentrations can be several micrograms per liter or greater (USEPA, 2001). The various sources of cadmium to the environment are shown in Figure 1. The emission ratio of anthropogenic to natural cadmium can be as high as 7:1.

Background concentrations of cadmium in waters range from <2 to 16 ng/L for pristine freshwaters where the cadmium cation predominates, and <5 to 100 ng/L in saline waters where cadmium is mainly complexed by chloride. Open seawater cadmium varies between 0.02 and 0.1 µg/L. The cadmium cation is the most bioavailable chemical form. Other factors affecting cadmium bioavailability in water include salinity, pH, dissolved organic carbon, and water hardness (calcium and magnesium). In sediments, humic material and acid volatile sulfides (FeS) will be important controls on cadmium bioavaiability. Other ions that can affect cadmium uptake or toxicity include manganese, zinc, and selenium.

In recent years, much concern has been expressed about the fate (transport, distribution, and transformation) of this potentially toxic metal in aquatic ecosystems.

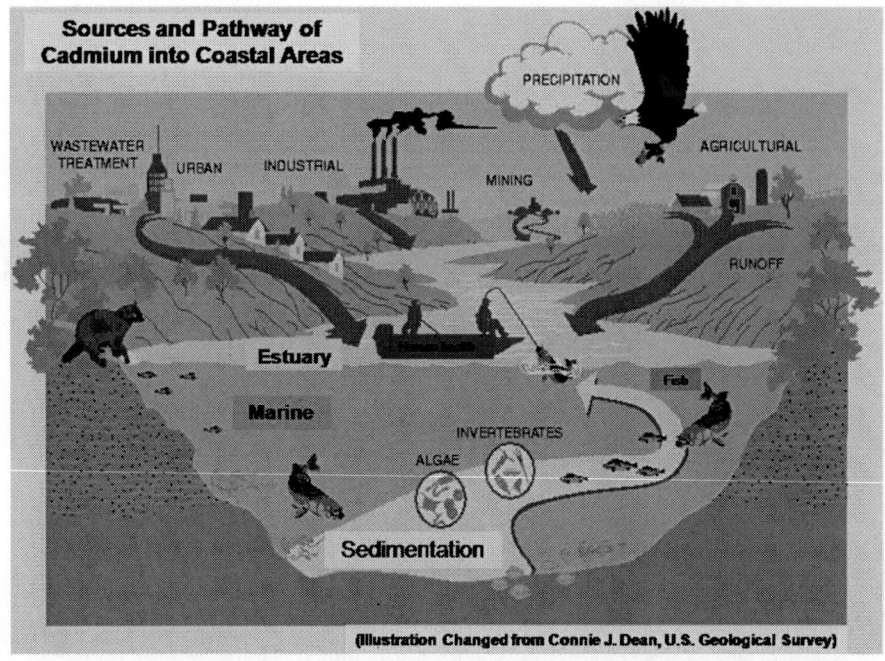

Figure 1. Different sources and pathway of cadmium into the aquatic environment.

2.1. NATURAL OCCURRENCES

Cadmium occurs in nature as a natural component of rocks, sediments, soils and dusts, air, water, and plant and animal tissues, where it appears to cause no harm either to human beings or to the environment. Its geochemical behavior is similar to that of zinc because of the similar electron structures and ionization potentials of the two elements.

The cadmium concentration found in rocks varies widely, ranging from 0.1 to 1 g/kg in soils, and to 100 g/kg in phosphatic rock. Cadmium forms an oxide and carbonate under natural conditions and is a constituent of both primary zinc minerals such as sphalerite and secondary minerals such as smithsonite. Because of its high mobility in the environment, large amounts of cadmium reach the water cycle, leading to extensive particle formation and subsequent sedimentation. Remobilization from sediments also occurs, enhanced by numerous chemical and biological factors, leading to the widespread dispersion of cadmium into ground water. Cadmium in marine sediments, range from 0.1 to 1.0 mg/kg in the Atlantic and Pacific oceans (Thornton, 1992).

Soils tend to reflect the chemical composition of the parent materials from which they were derived. The average cadmium content of surface soils ranges from 0.07 to 1.1 mg/kg in many parts of the world and all values over 0.5 mg/kg reflect anthropogenic inputs (Thornton, 1992).

Cadmium enters the atmosphere through natural processes such as weathering and volcanic emissions and is then deposited by precipitation into water bodies. In unpolluted and uninhabited areas, cadmium concentrations in the air are less than 1 ng/m^3.

Cadmium tends to accumulate in the leaves than in seed or root crops of plants that are grown on cadmium-containing-soils. Cadmium concentrations vary widely but generally range from less than 5 mg/kg to 10 mg/kg in most mammals. In oysters, mussels, amphipods, and squid, cadmium accumulates to concentrations of more than 1 mg/kg, whereas the concentrations in fish tend to be lower. Such low concentrations are due to cadmium sequestration by phytoplankton and/or the feeding behavior of invertebrate organisms (Lee et al., 2000). In humans, the main route of exposure to cadmium under normal conditions occurs through dietary ingestion and inhalation of tobacco smoke.

2.2. ANTHROPOGENIC SOURCES

World production and consumption of cadmium has continued to rise in most industrialized countries in recent years, principally as a by-product in zinc refining.

The end uses of cadmium in different industrial processes include production of television tube phosphors, preparation of special alloys and solders, metal plating, nuclear reactor shields and rods, pigments in yellow or brown paints (for coloring plastics, glass, and enamels), stabilizers for processing PVC polymers, nickel-cadmium rechargeable batteries, and electronic waste. Cadmium coating are mainly applied via electroplating or dipping to another metal as a thin film for protection against corrosion. Seldom is it possible to recover the metal economically. The use of cadmium in alkaline rechargeable batteries has potential environmental hazards in view of the amounts of nickel and cadmium involved. These uses lead to the flow of cadmium into the environment in the form of atmospheric emissions, liquid effluents and wastewaters, as well as sludges and solid wastes.

Chemical fertilizers of the super-phosphate group can contaminate the soil because they contain from 0.05 to 170 mg Cd/Kg. In addition to the above-mentioned sources, mining also provides at least two indirect sources of environmental cadmium. Firstly, the products of fossil fuel combustion in thermal power stations are the source of 50% of the atmospheric cadmium emissions. Secondly, cadmium that is associated with zinc ores does not remain contained within the localized mining region but rather seeps into local soils and drainage waters (Pinot et al., 2000). So anthropogenic sources, including smelter emissions and the application of fertilizers and sewage sludge to land may lead to contamination of both soils and crops. Besides incidental exposure through anthropogenic sources, tobacco smoke represents a major source of consistent and self-induced exposure to cadmium.

The majority people of the world live in town and cities and are subjected to exposure in the urban environment. The major anthropogenic sources of cadmium in the ambient air are non-ferrous metal smelters and uses of coal for combustion. Cadmium concentrations in the surrounding air depend on the population and urbanization of the region. Ambient air cadmium concentrations have been generally estimated as extremely low and may range from 0.1 to 5.0 ng/m^3 in rural areas, from 2.0 to 15.0 ng/m^3 in urban areas, and from 15.0 to 150 ng/m^3 in urban and industrialized areas, particularly near metal refining and processing plants (OECD 1994). In urban areas, the cadmium concentrations may be higher due to various urban, transportation and combustion activities.

In house dust, cadmium is approximately six times greater than the concentration in soil, which presumably originates from cadmium-containing materials such as abraded plastics and possibly paints, presenting a possible risk to young children, who may accidentally ingest it. In addition, cadmium may be introduced by urban and motorway dusts, all of which are especially important considering the increasing development in populated, urban environments (Culbard et al., 1988).

In summary, the main anthropogenic sources of cadmium in water relate to ore mining (including mining wastes and mine waters), metallurgical industries, industrial use, municipal sewage effluents, cadmium-contaminated phosphorus-containing fertilizers, the disposal and application of sewage sludges on land, and to the contaminated agricultural soils. Furthermore some researchers have shown that most cadmium additions to water or land are from atmospheric deposition (Nriagu and Pacyna, 1988). So these sources, together with atmospheric precipitation, are thought to constitute the critical pathway for the contamination of organisms and of humans through the food chain.

2.3. STATUS OF CADMIUM POLLUTION IN ESTUARINE AND COASTAL AREAS

Anthropogenic and airborne sources of cadmium entering the seas and oceans directly from atmospheric deposition mainly in particulate forms and, to a lesser extent, dissolved in rain and snow and indirect via river runoff.

Rivers are subjected to cadmium inputs from raw or treated sewage. River sediments generally reflect the neighbouring soils and mineral workings. High cadmium levels are invariably accompanied by high levels of other trace metals. As a result of these inputs, there are enhanced levels of cadmium in near-shore sediments and coastal sea-water. For example, the River Rhine has a range of 1 to 10 µg Cd/L. The general range for clean rivers and lakes is about 0.1 to 1.2 µg Cd/L, but 1 to 36 µg Cd/L in polluted industrial rivers in the United Kingdom, the USA, and in Europe. Values of 1 to 9 µg Cd/L were found for the Jintsu river in Japan which flows through the area where Itai-itai disease occurred and 30 µg Cd/L was found in the drainage streams of a nearby ore mine. In the United Kingdom, 3 to 95 µg Cd/L was found in streams in North Wales, while 5 to 20 µg Cd/L was found in areas in Cornwall affected by ore mining.

The transport of cadmium from freshwaters to the sea occurs either in particulate or soluble forms. The specific form depends on the state of the river, its mineralization and its sources of pollution, as well as on unidentified local factors. So quantification is difficult and studies on the transfer of the different forms of cadmium from freshwater to the sea are inconclusive. Certain estuaries are conservative in most trace metals. The daily inputs to inshore waters of the North Sea from the Humber estuary have been assessed as 37.15 kg Cd/d. The relative proportions and magnitudes of these inputs should be of general interest to other fairly heavily urbanized and industrialized estuary areas (UNEP, 1985 and literature cited by this report).

In China, heavy metals are continuing to be introduced to estuarine and coastal environment through rivers, runoff, and land-based point sources with the rapid industrialization and economic development. The Changjiang estuary is the largest in the western Pacific Ocean and the fifth largest in the world, and is located near Shanghai (the largest city of China). It receives a large amount of industrial discharges, urban waste disposals, and agricultural and domestic sewage effluent from the Changjiang delta area (the most economically developed area in China) and Changjiang River water. Figure 2 shows cadmium concentrations in various estuarine water from 13 sampling sites of Changjiang estuary, all sampling sites of which are located in latitude 30°54′N - 31°38′N, longitude

121°40′E - 122°30′E (see Figure 3). Depth profiles showed that the average cadmium concentrations of water column were 4.40 ± 2.20µg Cd/L (n=122, surface layer), 4.48 ± 2.05 µg Cd/L (n=118, middle layer), 5.02 ± 2.13µg Cd/L (n=118, bottom layer), respectively.

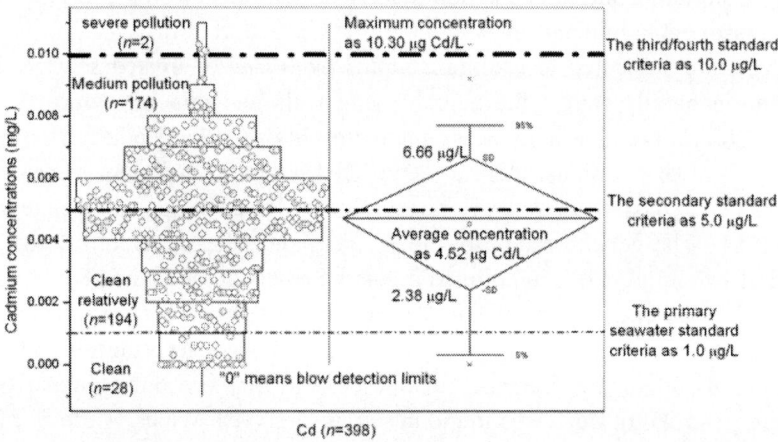

Figure 2. Distribution of cadmium concentration measured in Changjiang estuary and comparison with seawater standard criteria.

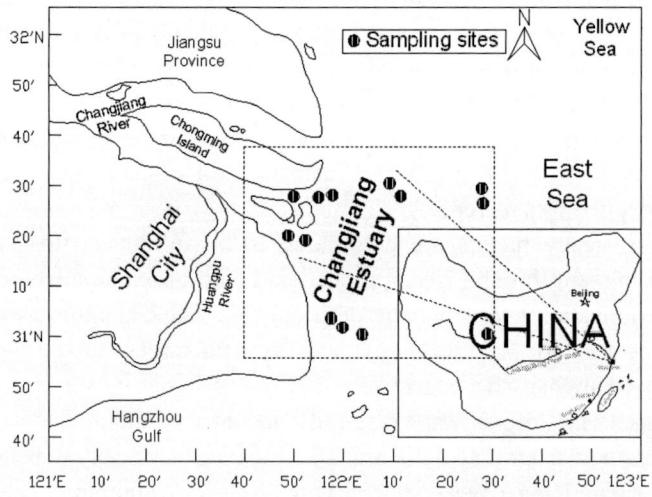

Figure 3. Locations of the sampling sites and schematic map of the Changjiang estuary, China.

Cadmium in aquatic ecosystems tends to accumulate in sediments, acting as a source for benthic biota and possibly for re-entering the water column. Measured cadmium concentrations of surface sediments varied from 0.11 to 1.11 mg/kg in Xiamen Bay (sampling sites shown in Figure 4). Taken together, these coastal and estuarine monitoring data suggest the presence of cadmium contamination and polluted situations in China[*] (Wang et al., 2009a).

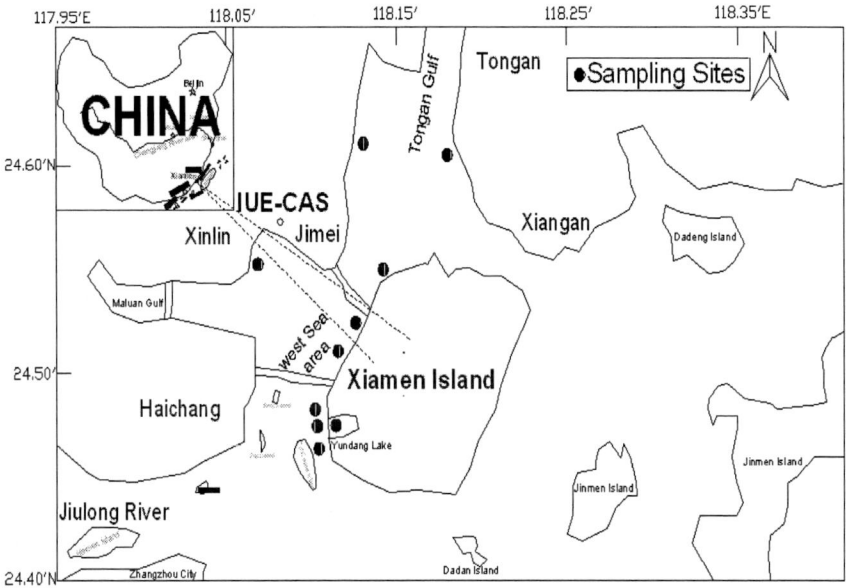

Figure 4. Locations of the sampling sites and schematic map of the Xiamen West Sea area, China.

The high cadmium exposures in estuarine and coastal areas of China are most likely related to human activities, such as industrial and municipal effluents, landfill leaching, non-point source runoff, and atmospheric deposition, although a direct link cannot be established with the available data.

---

[*] It was noteworthy that although the number of samples taken and analyzed was large, the total number of samples and sampling sites in the monitoring program may be limited, especially for surface water and sediment in saltwater systems, given the size and diversity of saltwater and estuarine ecosystems in Changjiang estuary and Xiamen Bay. So in future, monitoring of surface water and sediment should be consistent spatially and temporally, especially in those geographic locations where cadmium is widely applied and used.

2.4. SPECIATION AND BIOAVAILABILITY OF CADMIUM IN AQUATIC ENVIRONMENT

The free cadmium cation [Cd^{2+}] has been shown to be the most bioavailable and toxic form of the metal for many aquatic organisms, which means that its bioavailability and toxicity is diminished when cadmium becomes associated with particulate matter or forms of inorganic or organic complexes. Bioaccumulation of cadmium from the aquatic environment is controlled by many interrelated aspects of water chemistry, which involve cadmium speciation in the water and the ways in which the cadmium species compete with chemical ligands in the water, with binding sites on both sedimentary and suspended particulate matter, as well as on organisms (Figure 5).

In freshwater, dissolved cadmium is present mainly in free ionic form, although ionic cadmium tends to associate with colloidal and particulate matter, and the carbonate and the hydroxides in soft waters of which become increasing important at higher pH. The extent to which these occur and the factors (particularly pH dependent) that influence the association of cadmium with particles has biological significance. Dissolved cadmium is directly available for uptake by biota, and accumulation or toxicity of cadmium has similarly been related to Cd^{2+} in freshwater studies, although many heterotrophs ingest particles and assimilate material from them, and the distribution coefficient that relates particulate to dissolved cadmium for a wide range of natural waters is relatively constant. In seawater, cadmium chemistry is dominated by chloride complexation, so 97% of total cadmium may be in the form of chloride complexes. $CdCl^+$ and $CdCl_2^0$ predominate, although $CdCl_3^-$ is also present. A high proportion of cadmium is also associated with particles and is present as complexes in coastal and estuarine waters, (Mackay, 1983). Moreover, cadmium accumulation by marine animals and any resultant toxicity have often been shown to relate inversely to salinity (Wright and Welbourn, 1994).

Dissolved organic carbon (DOC) has the capacity to decrease the availability of cadmium by binding or complexing free cadmium, i.e., by changing cadmium speciation, bioavailability, and thus toxicity. Huebert and Shay (1992) indicated that cadmium bioavailability and toxicity in water may be greatly reduced by the addition of a strong complexing agent such as EDTA. Organic chelates such as humic acids are also likely to bind cadmium and appear antagonistic to the toxicity of cadmium. Although this binding is not as strong as EDTA, such studies of which implicated DOC in influencing cadmium bioavailability in the aquatic environment.

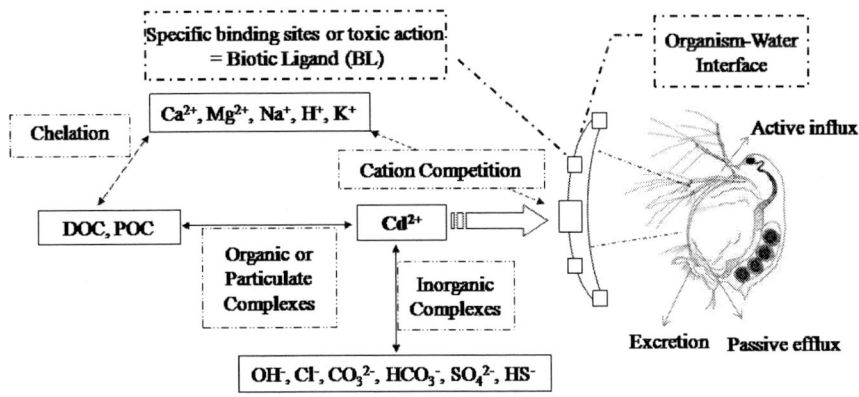

Figure 5. Conceptual diagram of cadmium speciation and cadmium-binding sites based on the biotic ligand model (BLM) showing inorganic and organic complexation in the water and interaction of metals and cations on the biotic ligand, adaptation to the zooplankton. (after Di Toro et al., 2001). POC=particulate organic carbon; DOC=dissolved organic carbon; CO_3^{2-}=carbonate; HCO_3^-=bicarbonate; OH^-=hydroxide; Cl^-=chloride; SO_4^{2-}=sulfate; HS^-=sulfide.

Many factors in aquatic environment might also affect the test results of the cadmium toxicity to aquatic organisms, for example, hardness (including calcium and manganese by influencing membrane integrity) and alkalinity are often thought of as having a major effect on the toxicity of cadmium in freshwater (Figure 6), although the observed effect may be due to one or more of a number of usually interrelated ions, such as hydroxide, carbonate, calcium, and magnesium.

In estuarine environments, where organisms are exposed to low and fluctuating salinities, cadmium speciation may be further complicated by other aspects of water chemistry, such as changes in organic carbon loading and significant alterations in the ratio of major to minor ions in the water body. In particular, the role of calcium in modifying cadmium availability may become increasingly important in fresher water.

In saline and also in freshwater environments, it is likely that the major effect of calcium on cadmium toxicity is through competition for organismal binding sites, rather than by altering its speciation. Calcium is antagonistic to cadmium uptake and release by the gill and may indirectly affect accumulation by other tissues.

The inverse relationship between calcium and cadmium is most easily explained as a competition for binding sites on the mussels (Figure 6).

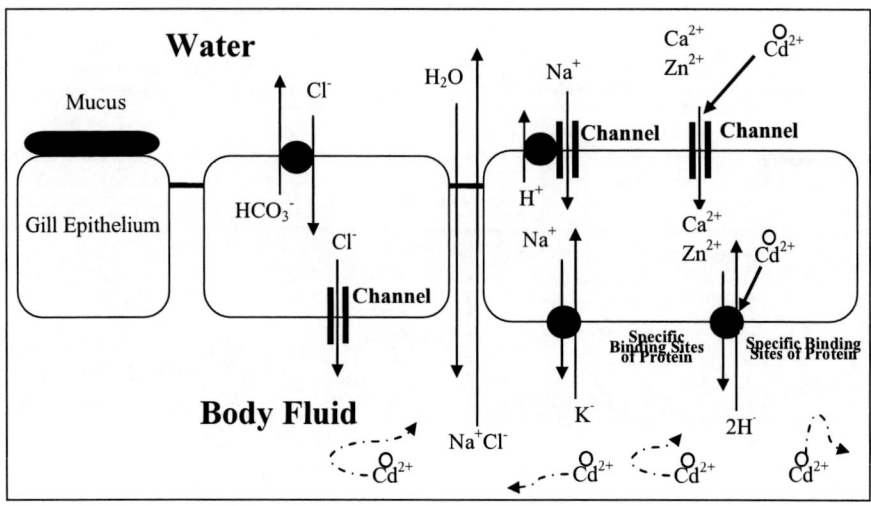

Figure 6. Schematic diagrams of transport pathways of essential ions (Na^+, Ca^{2+}, Zn^{2+} and Cl^-) and their interactions with toxic cadmium cations (Cd^{2+}) in the epithelium of aquatic organisms. After uptake (transport across the plasma membrane), free reactive Cd^{2+} will circulate through the body fluid and bind some proteins, then transport them into different compartments, where they are detoxified or not.

Furthermore metals such as zinc, iron and selenium can potentially influence the uptake and toxicity of cadmium by competing with cadmium for binding sites on epithelial and cell membranes of aquatic species or for other ligands that might affect cadmium availability indirectly, and hydrogen ion plays a role through the physiological processes and the cadmium speciation (Wright and Welbourn, 1994).

In the sedimentary environment, bioaccumulation of cadmium is largely dependent on soluble cadmium present in pore waters and may also depend upon the chemical forms of solid phases of cadmium, since the preponderance of available ligands, both organic and inorganic, has a major influence on cadmium availability (Nebeker et al. 1986). Despite the dominance of iron oxides in sequestering cadmium in aerobic sediments, humic acids have also been shown to possess significant cadmium binding capacity in freshwater sediments and may reduce the availability of cadmium to benthic organisms. Furthermore, cadmium is held partly in combination with carbonates and sulfides, and partly in complex organic combinations. In the latter example, cadmium combines with the sulfur-rich fractions of organic matter of which there is great excess, for the inorganic and organic contributions are widely variable. So the combined effects of pH, organic complexing agents, and redox potential of sediments need to be considered in assessing the availability of cadmium. The concentrations of acid-volatile sulfide have been advanced as a predictor of sediment cadmium toxicity based on the fact that any free Cd^{2+} in the pore waters will exchange with acid volatile sulfides (largely FeS) informing insoluble CdS (Di Toro et al., 1990).

The toxicity and impact of cadmium on aquatic organisms depends on different cadmium forms, which can have different toxicities and bio-concentration factors. In most well oxygenated freshwaters that are low in total organic carbon, $[Cd^{2+}]$ will be the predominant form. Precipitation by carbonate or hydroxide and formation of soluble complexes by chloride, sulfate, carbonate, and hydroxide should usually be of little importance. In saltwater with salinities from about 10 to 35‰, cadmium chloride complexes predominate. In both fresh and saltwater, particulate matter and dissolved organic material may bind a substantial portion of the cadmium, and under these conditions cadmium may not be bioavailable due to this binding. So the bioavailability of cadmium and, consequently, its accumulation in tissues depends on a number of factors.

Chapter 3

TOXICITY OF WATERBORNE CADMIUM TO SALTWATER AQUATIC ORGANISMS

Cadmium is a relatively rare element that is a minor nutrient for aquatic organisms at low concentrations, but is toxic to aquatic biota at concentrations only slightly higher. In aquatic environments, organisms are exposed to cadmium in dissolved and particulate-bound forms, including ambient water, sediments, and food (Figure 7).

Cadmium can be taken up by bacteria, phytoplankton, zooplankton and fish directly or through the food chain as a potentially toxic metal, which can enter the organism body via waterborne and dietary pathways, although regulatory assessments of metal toxicity to aquatic organisms assume that toxic effects are caused by dissolved metals.

In recent years, concern has been expressed about the possible effects on aquatic organisms from exposure to cadmium. Both acute and chronic toxicity tests are essential for better understanding of the response of aquatic organisms to cadmium. This chapter provides many different studies on cadmium toxicity to aquatic biota. A partial listing of recent data to illustrate the range of cadmium concentrations eliciting toxic responses is provided in Figure 8 for acute cadmium toxicity in aquatic biota and a broad spectrum of sublethal effects, respectively

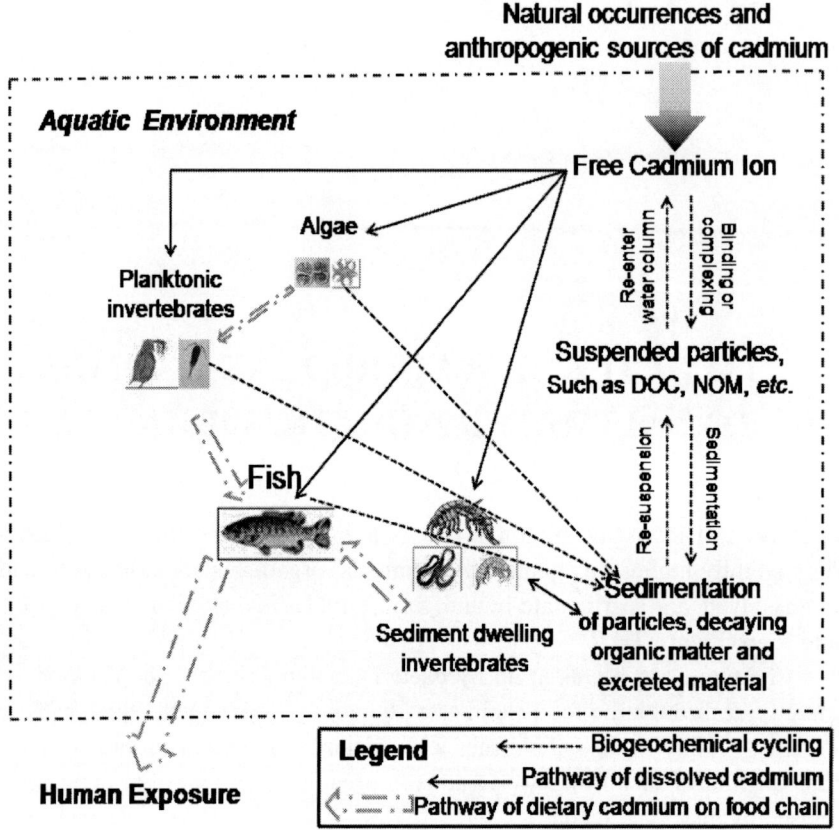

Figure 7. The possible routes of aquatic organism exposure to cadmium (direct uptake from water phase or through food uptake).

3.1. ACUTE TOXICITY

The acute toxicity of cadmium to saltwater organisms has been determined for 48 species of invertebrates and 12 species of fish, representing the required different taxonomic families (Figure 8). Having data for a variety of organisms is helpful, because different test organisms provide different types of information about the toxicity of cadmium. The range of species tested not only included marine Arthropoda, Crustacea ranging from copepods (*Acartia tonsa*) to amphipod (*Corophium insidiosum*), to mysid (*Americamysis bahia*), to crabs (*Cancer magister*), and shrimp (*Crangon septemspinosa*), but included several

Mollusca (bivalve), filter-feeding mussels (*Mytilus edulis*), the Pacific oyster (*Crassostrea gigas*), and the deposit-feeding clam (*Mya arenaria*), as well as the squid (*Loligo opalescens*), and mud snail (*Nassarius obsoletus*). It also included Chordata (Osteichthyes), Annelida (Polychaeta), Echinodermata (Echinoidea).

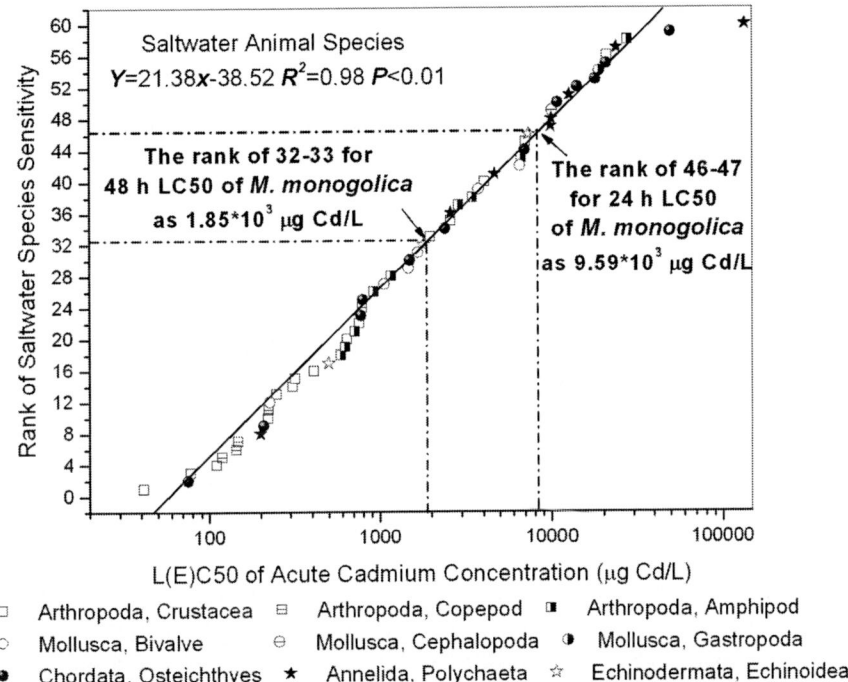

Figure 8. Distribution of acute cadmium toxicity data from various saltwater organisms of different phyla or class; Arrows show how 24- and 48-h LC50s of *M. monogolica* Daday (Saltwater zooplankton species from China) exposed to cadmium compare and rank with these data (USEPA, 2001).

As indicated by the data (Figure 8), SMAVs (Species Mean Acute Values) or LC50s (Mean Lethal Concentrations) of saltwater cadmium varied by 5 orders of magnitude, the highest concentrations being for the oligochaete worm, *Monopylephorus cuticulatus* (135 mg Cd/L) and the lowest 41.3 µg Cd/L for the mysid, *Americamysis bahia*. Ninety-four percent of the data fell in the three decades between 100 and 100,000 µg/L.

For saltwater polychaetes, the acute values range from 200 µg Cd/L for *Capitella capitata* to 135,000 µg Cd/L for *Monopylephorus cuticulatus*, and for saltwater molluscs, SMAVs range from 228 µg Cd/L for the Pacific oyster

(*Crassostrea gigas*), to 19,170 μg Cd/L for the mud snail, *Nassarius obsoletus*. For 12 fish species, SMAVs range from 75.0 μg Cd/L for striped bass, *Morone saxatilis*, to 50 mg Cd/L for sheepshead minnow, *Cyprinodon variegates*.

It is noteworthy that the rank of *M. mongolica* (an indigenous or native species of China) for species sensitivity in Figure 8 was in the range of 46-47 and 32-33 for 24- and 48-h LC50 values, respectively, which indicates that *M. mongolica* is moderately sensitive to cadmium stress compared with different phyla or different orders of the same phylum (Arthropoda) and class (Crustacea), including amphipod, copepod, mysid, decapods (shrimp, crab), and isopod.

Data given in Figure 8 also indicate that cadmium is not very toxic over short exposure periods, and the LC50s for a wide range of species are usually in excess of 1 mg Cd/L[*]. It is most unlikely that these concentrations will occur in the future, even in the most polluted situations of Figure 3.

3.2. SENSITIVITY OF SPECIES

Both invertebrate and fish species display a wide range of sensitivities to cadmium in Figure 8. Of the 61 saltwater genera for which acute values are available, The SMAVs for saltwater invertebrate species range from 41.29 μg/L for a mysid to 135,000 μg/L for an oligochaete worm. So the most sensitive, Mysid, *Americamysis bahia*, is 3,270 times more sensitive than the most resistant, Oligochaete worm, *Monopylephorus cuticulatus*.

The average rank number of different class or species is shown in Table 1, which indicated that saltwater aquatic animals had a broad range of cadmium tolerance[*] including Anthropoda, Echinodermata , Mollusca, Chordata and Annelida.

---

[*] Generally, saltwater species were more resistant to cadmium than freshwater species. The reason was that the acute toxicity of cadmium was related to salinity and hardness, which influence the cadmium bioavailability, i.e., greater cadmium toxicity in low salinity versus high salinity environments. The acute toxicity of cadmium generally increases as salinity decreases. But regrettably, there was no consistent salinity-toxicity trend observed for the data, and a salinity correction factor was not attempted at the present time.

[*] Tolerance and resistance are genetically controlled properties, but while tolerance refers to the existence within a species of a subgroup that is able to function under higher exposure to a toxic substance than the normal members of the species, resistance refers to an entire species being able to function at elevated levels of an otherwise toxic substance. It should be noted that the distinction could break down over time; one could anticipate that as the tolerant subgroup diverged from the parent species, eventually speciation might occur and the tolerant form would then be resistant. For the purposes of the present chapter, we have not attempted to adhere to the

Anthropoda, especially copepods, appeared to be the most sensitive in this group (Table 1 and Figure 8), but Annelida were the most resistant to cadmium, and no effects were recorded at concentrations less than 100 μg Cd/L. Echinodermata also appears to be very resistant to cadmium, and no harmful effects were recorded at concentrations less than 100 μg Cd/L (UNEP, 1982; USEPA, 2001).

Table 1. Rank number of different class or species based on toxicity according to Figure 8.

Phylum		Species rank number	Average rank number	
			Class	Species
Anthropoda	Copepod	5,6,7,11,24	10.6 (n=5)	22.76(n=29)
	Amphipod	18,19,21,26,28, 37,38,43,58	31.11 (n=9)	
	Other Crustacea	1,3,4,10,13,14,15,16, 20,22,33,35,40,45,56	21.8 (n=15)	
Mollusca	Bivalve	12,27,29,31,39,42	30 (n=6)	35.375(n=8)
	Cephalopoda	49	49(n=1)	
	Gastropoda	54	54(n=1)	
Echinodermata, Echinoidea		17,32,46,	31.67(n=3)	
Chordata, Osteichthyes		2,9,23,25,30,34, 44,50,52,53,55,59	39.64(n=12)	
Annelida, Polychaeta		8,36,41,47,48,51,57,60	43.5(n=8)	

Organisms are able to survive in polluted environments of high cadmium concentrations because the processes of detoxification must have been developed.

3.3. CHRONIC TOXICITY

Acute cadmium toxicity occurs over a wide range of concentrations depending on the organism, the chemical conditions, and the type of test, as do chronic effects. In attempting to elucidate the actual models of toxic action for

---

particular distinction. Rather we have addressed biochemical or physiological mechanisms of tolerance/resistance.

cadmium, it is perhaps most instructive to consider initially low-level, long-term (i.e., chronic) studies.

Some researchers conducted a 23-day life-cycle test with the saltwater invertebrate, Mysid, *Americamysis bahia* at 20 to 28 °C and salinity of 15 to 23 g/kg in three chronic cadmium concentrations. Survival was 10% at 10.6 µg/L, 84% at the next lower test concentration of 6.4 µg Cd/L, and 95% in the controls. No unacceptable effects were observed at 6.4 µg Cd/L, therefore the chronic toxicity limits are 6.4 and 10.6 µg Cd/L, with a chronic value of 8.24 µg Cd/L (USEPA, 2001).

The effect of cadmium on the reproduction strategy of saltwater zooplankton, the cladoceran *M. monogolica* was investigated by Wang et al. (2009a). After a 21-day exposure of the 18 ± 6-h old neonates to 0 (control), 4.1, 12.9, 26.7, 47.1, 67.89, 91.6 and 171.8 µg Cd/L at 20 ± 0.5 °C, the authors compared the survival, number of neonates per female, first day of reproduction and neonate size of the cadmium exposures to the controls. Cadmium negatively influences both survivorship and fecundity of several genera including *Moina monogolica* Daday. However, the effect on fecundity may not be with the same magnitude on survivorship. For example, cadmium at 4.1 µg Cd/L affected the fecundity of *M. monogolica* but had no effect on the survivorship. The reproductive NOEC (no-observed-effect concentration) of 21-day static-renewal test was 1.11 µg Cd/L cadmium, and the reproductive LOEC (lowest-observed-effect concentration) was 3.01 µg Cd/L. The resultant chronic value was 1.78 µg Cd/L (Wang et al., 2009a).

Zooplankton, particularly rotifers and cladocerans are useful for chronic tests since they are not only sensitive to cadmium stress, but also due to their wide distribution, are easy to maintain under laboratory conditions, their parthenogenetic life cycle ensuring the supply of several individuals with little genetic variability and relatively higher population growth rates. Among several biological variables, population growth characteristics serve as sensitive indicators of toxic stress in zooplankton (Garcia et al., 2004).

Population growth studies are helpful to quantify sublethal effects of toxicants including cadmium to zooplankton because small changes in survivorship and fecundity are eventually summed up in peak abundances and growth rates (Halbach et al., 1984). For this, single species tests are appropriate because they yield information both rapidly and quantitatively to evaluate the direct effects of toxicants which could be extrapolated to natural conditions with some precision.

3.4. BIOACCUMULATION (INTERNAL METAL EXPOSURE)

Cadmium metal is slightly soluble in water, although its chloride and sulfate salts are freely soluble. Particulate matter and dissolved organic matter may bind a substantial portion of cadmium in both freshwater and saltwater. Cadmium does not easily degrade in aquatic systems and tends to bind to sediments, but is also readily bio-accumulated by aquatic organisms. Bio-concentration factors (BCFs) in freshwater fish were as high as 12,400 and for a saltwater polychaete were as high as 3,160 (USEPA, 2001).

The BCFs for a saltwater fish were 48 from a 21-day exposure of the mummichog. However, among ten species of invertebrates, the BCFs range from 22 to 3,160 for whole body and from 5 to 2,040 for muscle. BCFs for five species of saltwater crustaceans range from 22 to 307 for whole body and from 5 to 25 for muscle, and whole-body BCFs for two species of grass shrimp, *Palaemonetes pugio* and *P. vulgaris* were reported as 203 and 307. The highest BCF was reported for the polychaete, *Ophryotrocha diadema*, which was attained a BCF of 3,160 after 64-days exposure using the renewal technique, although tissue residues had not reached steady-state (USEPA, 2001).

BCFs for four species of saltwater bivalve molluscs range from 113 for the blue mussel to 2,150 for the eastern oyster. In addition, the range of reported BCFs is rather large for some individual species. BCFs for the oyster include 149 and 677, as well as 1,220, 1,830 and 2,150. Similarly, two studies with the bay scallop resulted in BCFs of 168 and 2,040 and three studies with the blue mussel reported BCFs of 113, 306, and 710. The importance of metal speciation on cadmium accumulation in the soft tissues of *Mytilus edulis* was studied, where cadmium complexed as Cd-EDTA, Cd-alginate, Cd-humate, and Cd-pectate was bioconcentrated at twice the rate of inorganic cadmium. Because bivalve molluscs usually do not reach steady-state, comparisons between species may be difficult and the length of exposure may be the major determinant in the size of the BCF (USEPA, 2001).

Bioaccumulation is often a good integrative indicator of the organism exposures to metal in polluted ecosystems, but varies widely among metals, species, and environmental conditions, which could well be predicted by the dynamic multi-pathway bioaccumulation model (DYMBAM), capturing the biologically driven patterns that differentiate bioaccumulation among species and considering geochemical influences, biological differences, and differences among metals. It was noteworthy that the links between bioaccumulation and toxicity are complex, because bio-accumulated metal is not necessarily toxic. Toxicity is determined by the uptake of metal internally and the species-specific partitioning of accumulated metal between metabolically active and detoxified forms (Luoma and Rainbow, 2005).

3.5. MECHANISMS OF CADMIUM TOXICITY

The eco-toxicology of cadmium has put greater emphasis on gathering data on the sensitivity of species rather than understanding the reasons why one species is apparently more sensitive to cadmium than another, which drew the attention of researchers to the relationship between the potential toxicity of cadmium and to the risks presented by its accumulation in biota.

3.5.1. Metal Accumulation Patterns

Aquatic organisms can take up or accumulate trace metals in their tissues, via the surrounding aquatic environments or diet, based on different metal accumulation patterns, depending on whether these metals are essential to metabolism or not. The bio-accumulated concentrations are divided into two components—a metabolically available form and a stored detoxified form (Figure 9).

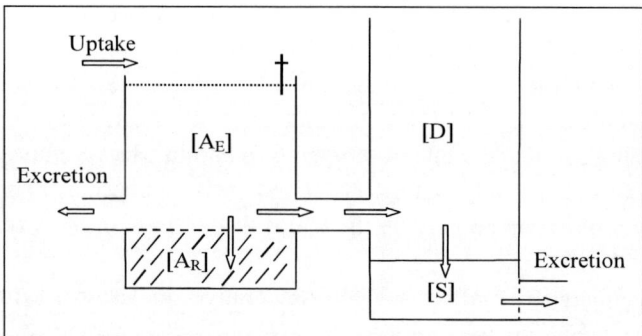

Figure 9. Generalized scheme showing the various compartments in which metals may be present and accumulate inside aquatic organisms (modified from Vijver et al., (2004)), [A_R] is metabolically metal pool, for nonessential metals this pool is negligible; [A_E] is excess pool above metabolic requirements, causing toxicity and eventually mortality when elimination or detoxification fluxes are slower than uptake rate; [S] is storage; and [D] is detoxification.

Metals in the detoxified form can be excreted or bound to proteins or other particular molecules of high affinity. Cadmium exist in a variety of bound forms including being associated with metallothioneins (MTs) and/or granules (Rainbow, 2002).

Essential metals may be subject to regulation either by limiting metal uptake at the level of the total body concentration or by involving organism-specific accumulation patterns with active excretion from the metal excess pool and/or storage in an inert form and/or excretion of stored (detoxified) metal (Figure 9) (Vijver et al., 2005). For non-essential metals, excretion from the metal excess pool and internal storage without elimination are the major patterns of accumulation, and body concentrations generally increase with increasing external concentrations because there are no excretion and no essential requirements. On uptake, waterborne cadmium will be in a metabolically available forms and will need to be detoxified if the amount of metabolically available cadmium in the body is not to exceed the threshold for toxic effects. Most accumulated metal might be expected to be in detoxified form and toxic effects are instigated when [U] exceeds [D] (Figure 10) (Rainbow, 2002).

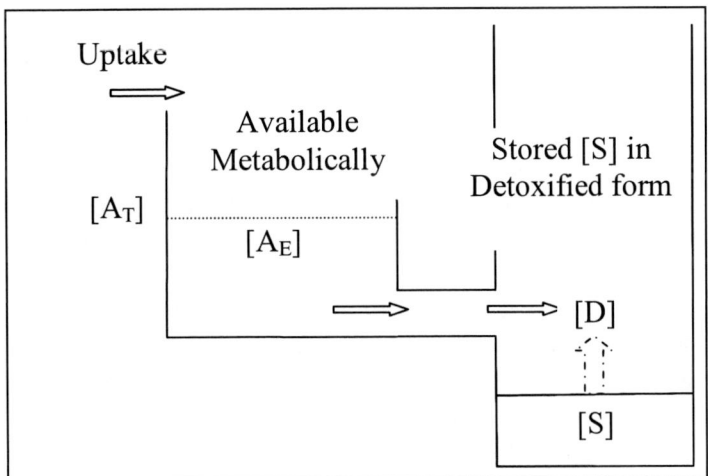

Figure 10. Cadmium accumulation pattern of aquatic organisms from waterborne sources showing net accumulation of cadmium without excretion of any cadmium taken up (modified from Rainbow, (2002)), [A_E]—excess cadmium over and above metabolic requirement, [A_T]—threshold concentration of metabolically available cadmium, above which the accumulated cadmium is toxic, [D]—metabolically available cadmium in excess of requirements is detoxified, [S]—the detoxified component of accumulated cadmium with no upper concentration limit.

Toxicity is related to a threshold concentration of metabolically available metal rather than to total accumulated metal concentration. The onset of toxic effects depends only on the concentration of accumulated metal in a metabolically available form. It follows, therefore, that toxicity occurs when the rate of metal uptake into the body exceeds the combined rate of excretion plus detoxification of metabolically available form. Organisms are able to control metal concentrations in certain tissues of their body to minimize damage of reactive forms of trace metals and to control selective utilization of essential metals (Vijver et al., 2005), but once the metabolically available forms passed a threshold concentration, the organism will then suffer toxic effects, initially sublethal but eventually lethal.

3.5.2. The Role of Metal-Binding Proteins, Metallothioneins (MTs)

Metallothioneins (MTs) are heat-stable, metal-binding proteins of low molecular weight (ca. 7 kilodaltons), high cysteine content (ca. 30%), and no aromatic amino acids. They occur throughout the animal kingdom as well as in plants, and are located mainly in the cytosol or in the nucleus, the amounts of which were dependent upon the tissue metal concentration. Two or more isomers of MTs may even exist in the same animal. Affinity for MTs is metal-dependent and correlated with the distribution of metal binding sites on the MTs as well as the stoichiometry of the different types of MTs. These differences in binding strength are relevant for the involvement of MTs in metal-metal interactions.

MTs can be induced by exposure to elevated levels of Hg, Cu, Zn and Cd, and have a high affinity for these metals, although transition metals have differing affinities for MTs (in decreasing order, Hg>Cu>Cd>Zn) and displacement reactions may occur dependent on the relative concentrations of these metals (Engel and Brouwer, 1987). All MTs may be induced as a result of cadmium exposure and are capable of binding cadmium. Detoxification often involvs binding to proteins such as MTs or forming insoluble metaliferous granules. Some MTs may be basally expressed and play a role in copper and zinc metabolism and regulation (Engel and Brouwer, 1987).

MTs play a dual role in most aquatic organisms. For essential metals, MTs may be important in controlling metabolically available concentrations by binding the metals in a nontoxic unavailable form until they are required for various metabolic processes, although regulatory and/or detoxification functions may overlap. Furthermore, their function has probably pre-adapted them to act as regulators in a detoxifying role for excess concentrations of copper and zinc (Engel and Brouwer, 1987). For non-essential metals, there is no evidence that

MTs have any cadmium regulatory capacity as determined by tissue or body concentration. Cadmium detoxification relies heavily on the ability of induction and binding to MT and subsequent available sequestration capacity, which was followed within the context of cadmium uptake dynamics and distribution within organisms and correlated with increased tolerance to this metal. MT induction can not be considered as an overriding principal of cadmium tolerance, which has several other modifying factors. For example, cadmium sensitivity may have as much to do with metal competition and the capacity of MT production, if the spillover theory is invoked as a reasonable basis for delineating tolerance (Wright and Welbourn, 1994).

3.5.3. Critical Body Residue (CBR)

Similar to the external concentration, only a portion of total metal body burden is biologically available for interaction with sites of toxic action. Toxicity is related to the metabolically reactive pool rather than to the total internal metal burden, the capacity of which has an upper limit (Figure 10). As the basis of internal metal sequestration over different organs and tissues, it is necessary to investigate subcellular metal partitioning for better understanding mechanisms of accumulation and toxicity using a fractionation procedure, which can isolate metal-rich granules and tissue fragments from intracellular and cytosolic fractions.

Metal-rich granules (MRG) and tissue fragments can be separated from intracellular (nuclear, mitochondrial, and microsomal) and cytosolic fractions (i.e., MTs and heat-sensitive proteins) through the fractionation method of different centrifugation steps (Wallace et al., 2003). The fractionation constituents reflect the metal-binding preferences by different ligands, and the underlying principle is that metals entering the body in a reactive form are first captured by reversible (labile) binding to proteins and other ligands, followed by localization to targets that have a stronger affinity for metals.

The amount of metal compartmentalized in cytosolic (such as organelles and microsomes), heat denaturated protein (enzymes), and tissue fractions (such as nuclei, cell membranes, tissue and intact cells) seems to be most indicative of toxic pressure, which is supposed to be indicative of the excess metal fraction shown in Figure 10. The fractions such as granules, lysosomes and heat stable proteins (MTs) are maybe more important for uptake, elimination or tropic transfer than for toxic action.

The Critical Body Residue (CBR) is defined as the threshold concentration of a substance in an organism that marks the transition between no effect and adverse effect, based on assumption that the total body concentration is proportional to the concentration at the target or receptor, the effect is proportional to the concentration of metal bound to the target site, and that this target site (biotic ligand) is in direct contact with the external (aquatic) environment, which integrates internal transport and metabolism processes and toxicity at specific sites of toxic action, and provides a better understanding of the internal compartmentalization of metals in organisms and its consequences for toxicity (Vijver et al., 2005).

Cadmium ions in excess of metabolic requirements and storage capacity are potentially toxic and must be removed from the vicinity of important biological molecules, although organisms can minimize accumulation of reactive metal species at the cellular level. Therefore, the capacity of internal sequestration has a huge impact on the sensitivity of an organism to cadmium. At least two major types of cellular sequestration can occur after increased exposure to cadmium and affect their toxicokinetic availability to organisms (Table 2).

The first mechanism of cellular sequestration involves forming distinct inclusion bodies of metal accumulation, such as granule types originating from the lysosomal system and containing mainly acid phosphatase and accumulating cadmium identified by x-ray microanalysis. The second type preventing cadmium toxicity is a cytoplasmic process involving a specific metal-binding protein (heat-stable proteins, MT), because binding to MT generally can reduce the availability of toxic metals.

Table 2. Various cellular cadmium species in an organism after partitioning the total internal burden

Cadmium species	Example
Free ionic form or complexed ion species	Cd^{2+}, $CdCl_2$, $CdCl^+$, $CdCl_3^-$
Bound to low molecular weight and functional proteins	hemoglobin, hemocyanine finger proteins
Bound in the active center of enzymes	cytochromes, carbonic anhydrase, superoxide dismutase
Bound to low molecular weight organic acids	citrate
Bound to metallothionein, to transport proteins or other sequestration proteins	ferritin
Bound in vesicles of the lysosomal system	intracellular granules
Precipitated in extracellular granules, mineral deposits, residual bodies, and exoskeletons	
Bound to cellular constituents potentially causing dysfunction	enzymes, ion channels, DNA

3.5.4. Links to Toxicity

Relationships between bioaccumulation and adverse effects are complex when different species are compared in different environments for different metals. Although sublethal toxic effects such as reproductive impairment could be observed when bio-accumulated metals were coincident with an increased concentration. Defining adverse ecological effects, several aspects should be considered that link bioaccumulation (internal metal exposure), internal metal reactions, and the hierarchy of biological responses. Accumulation of bioactive metal is more important than total bioaccumulation, and understanding toxicity requires consideration of more than just total metal accumulated in tissues.

Taking the biotic ligand model (BLM) theory as an example, which mainly linked metal speciation in solution with the amount of metal bound at the gill surface of fish, based on the fact that such binding involves competition between metals binding to the gill binding sites (biotic ligand) and other complexing agents in solution as well as competition for the sites by hydrogen ions, calcium and magnesium. The proportion of binding sites on the gill that are occupied by metal determines the degree of disturbance to ion regulation (Figure 6). Adverse effects are determined not only by the amount of metal that is bio-accumulated, but the toxic "action" at the target site(s). For organisms other than fish, the theoretical basis is that total body residues directly determine the effect, which in organisms is assumed to be directly related to the amount of metal bound to the external body surface. The toxic effects of tested organisms can be predicted at the moment, but is restricted to some small organisms that quickly achieve an internal equilibrium for the metals between body surface and environments (Rainbow, 2002).

The sub-cellular partitioning model (SPM) taking into account the cellular fates of metals was proposed to predict metal toxicity in aquatic organisms. It considers how metal accumulation and subsequent redistribution are directly related to metal toxicity and takes advantage of a recently introduced conceptual model that groups individual subcellular fractions (e.g., metal-rich granules [MRG], cellular debris, organelles, heat-denatured proteins [HDP], and heat stable protein [HSP]) into ecotoxicological relevant compartments. For instance, organelles and HDP are grouped as the metal-sensitive fraction (MSF), and HSP and MRG are grouped as the biologically detoxified metal (BDM). The correlations between metal toxicity and subcellular metal distribution of several aquatic organisms (e.g., phytoplankton, bivalves, and fish) have been tested (Miao and Wang, 2006, Wang and Wang, 2008).

The subcellular fractionation approach may be more useful in explaining the sublethal (chronic) cadmium toxicity than the acute toxicity. When the toxicity results are applied to field situations, it is expected that they should be sufficiently conservative to protect most species (e.g., 95%).

3.5.5. Species-Specific Effects

The physicochemical properties of cadmium and the physiology of the organism both influence cadmium uptake, distribution, tissue accumulation, and excretion, although a diversity of specific cadmium accumulation strategies is known at the level of organs and tissues in an organism, tissue- and organ-specific. Toxicity is ultimately determined by cellular mechanisms of cadmium accumulation (Vijver et al., 2005).

Cadmium accumulation patterns are variable among species, which include regulation of body concentrations of cadmium by some species, and vastly different concentrations among species and environments. The compartmentalization or sequestration of cadmium by invertebrates is also dependent upon many factors, such as organism life history, and cadmium pre-exposure, so the modes of toxic action that sequester the bio-accumulated internal cadmium concentration are species-specific. In summary, bioaccumulation varies widely among taxa, often reflecting basic differences in biology.

Crustaceans

In crustaceans, soluble cadmium is probably largely accumulated through the gills, where it may reach high concentrations and result in tissue damage. No evidence showed that crustaceans were able to regulate their body cadmium concentration relative to ambient levels, although copper and zinc were substantially regulated. Although some crustacean MTs may have a specific copper regulatory function, all have very reactive cadmium binding sites and even, more labile of cadmium loci may be present in some additional cases. In some instances this binding capacity may result in the accumulation of very high cadmium concentrations in the hepatopancreas of crustaceans, which generally contains between two and four MT isomers.

Much of the cadmium accumulated by aquatic invertebrates is bound to MTs in the cytosol of the organ predominantly used for accumulated cadmium storage. In certain circumstances of severe cadmium exposure, there is indirect evidence that cadmium from MTs may be deposited in an insoluble form in lysosomal residual bodies. If the cells containing these lysosomal residual bodies line a tract with external access, then there is the potential for any cadmium-rich cell inclusions to be excreted, offering potential for the final accumulation pattern to be identified. The exoskeleton also has been shown to be a significant site of cadmium deposition in crustaceans, although the degree to which this occurs via the hemolymph or through external adsorption remains open to question (Wright and Welbourn, 1994).

In the shore Green crab *Carcinus maenas*, cadmium becomes associated with haemolymph proteins, but is rapidly turned over to the hepatopancreas where it may be stored as inorganic granules or associated with MT in proportions that depend on the degree of cadmium exposure.

Molluscs

In the bivalve molluscs, cadmium is principally accumulated via the gill from water having low environmental cadmium concentrations and can be rapidly bound into a non-toxic complex of mollusc tissues such as MTs, but retained within the body and excreted very slowly. Harmful effects may follow when the cadmium binding sites are saturated, because accumulation occurs in tissues which may be sites of toxic action (Wright and Welbourn, 1994).

Within the gill, the role of granular amoebocytes in cadmium binding has been clarified. Some researchers suggested that the cadmium-binding proteins in amoebocytes of cadmium-exposed oysters (*Crassostrea gigas*) were present and MTs in the cytosol of *Crassostrea virginica* gills that progressively sequestered cadmium at the expense of the granular fraction was located.

As for the marine mussel *Mytilus edulis*, kidney seems to be a major target organ for cadmium, where it may be sequestered in lysosomes resulting in granular concretions. When the kidney tissue of Bay scallop (*Argopecten irradians*) was exposed for 5 days to 700 µg Cd/L, the concretions or granules contained 60% of the accumulated cadmium. However when *Mytilus edulis* was exposed to 100 µg Cd/L for 3 months, even 85% of the cadmium in membrane-limited granular structures may have been associated with MTs, sulfur and sometimes phosphorus in membrane-bound vesicles. The cadmium excretion rate in *Mytilus edulis* was 18 times slower than the uptake rate and 50% of accumulated cadmium in *Crassostrea virginica* was lost in 60 days in clean water. Mussels loaded with 564 mg Cd/kg dry weight lost 47 mg/kg in a 42-day depuration period, in which time the fraction bound to MT rose from 22 to 78%.

Other researchers, also, found that the cadmium half-life in *Saccostrea echinata* was very long.

The granular fraction of scallops declined following three week exposure as the metal became preferentially bound to intracellular thiol groups (MTs), and high concentrations of cadmium have been found in the digestive gland bound to both high molecular weight proteins and MTs.

Fish

In fish, tissue cadmium accumulation is largely associated with MTs binding, and in contrast to invertebrates, granular concretions seem to play an insignificant role in cadmium sequestration. Cadmium has been shown to affect several enzyme systems, including those involved with neurotransmission, trans-epithelial transport, intermediary metabolism (sometimes in vitro) exposures to high cadmium concentrations.

For plaice (*Pleuronectes platessa*) and dabs (*Limanda limanda*), cadmium accumulates in the liver and gills rather than in their muscle and liver concentrations began to increase only after 70 days exposure to 5 μg Cd/L. ^{109}Cd pulse experiments indicate a biphasic transfer of cadmium from gills to kidney and liver, where there is little subsequent release. In minnows (*Phoxinus phoxinus*) and other species, cadmium was rapidly accumulated in the head region from water by the gills, probably resulted from localization in the olfactory mucosa, and the gut mucosa did not appear to be a major site of cadmium uptake from water, although this may change according to diet. However in several species of fish, the gut contained "intestinal corpuscles" - a mixture of mucous cells, mucous, and granules - which had a high cadmium concentration.

In the stone loach *Noemacheilus barbatulus*, cadmium sensitivity decreased as a result of higher availability of binding sites on existing MTs, compared with the rainbow trout *Salmo gairdneri*, where cadmium binding to existing and induced MTs was limited by relative cadmium, zinc, and copper concentrations in several tissues, particularly the liver. When plaice was injected with a cadmium dose of 100 μg/kg, MT induction exceeded the sequestration capacity of the induced MT and led to reduced production of the enzyme ethoxyresorufin o-deethylase (EROD) and of MT itself (Wright and Welbourn, 1994).

Polychaeta

In the polychaete *Neanthes arenaceodentata*, MTs had no capacity to perform any homeostatic function with respect to cadmium, the binding of which seems to be a passive function of the number and affinity of available binding sites, although this species apparently possesses some capacity to adjust cadmium uptake and efflux to reduce long-term accumulation.

Chapter 4

TOXICITY OF DIETARY CADMIUM TO AQUATIC ORGANISMS

Regulatory assessments of cadmium toxicity are originally mostly based on dissolved cadmium concentrations with the assumption that toxicity is caused by dissolved (or waterborne) cadmium only, and does not include the possible impact of cadmium associated with the particulate phase (e.g., phytoplankton, suspended particulate matter, sediment; food particles = dietary exposure). In recent years, more and more studies have shown that dietary cadmium uptake and accumulation is possible and very important (Wang et al., 2007; Wang and Fisher, 1999). Although situations exist where dissolved uptake alone describes most of bioaccumulation, the correlation between model forecasts and observations would be weak overall if only dissolved uptake was considered. Cadmium bioaccumulation from dissolved and dietary sources was 50% of total cadmium bioaccumulation separately in the bivalves *Mytilus edulis* and *Mytilus balthica*, and bioaccumulation of cadmium in nature agreed well with the forecasts when waterborne and dietary routes of uptake were considered. Assimilation efficiency of cadmium from food is much greater than that from water in *Daphnia magna* (Geffard et al., 2008). All of which suggested that dietary metals could play a crucial role in their health as strong evidence, although the toxicity of dietary metals on primary consumers was need to be understood further.

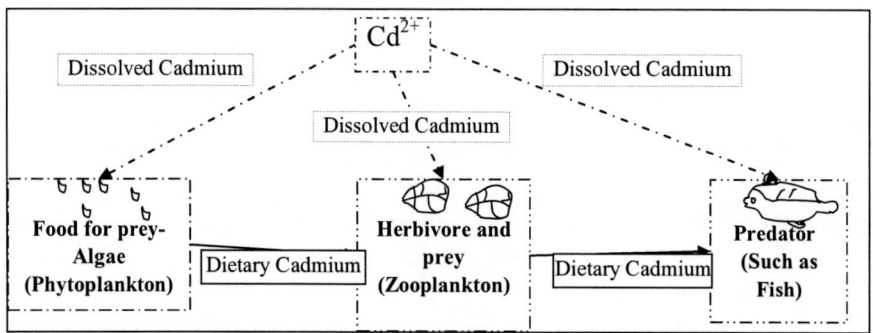

Figure 11. Diagrammatic representations of various aquatic organisms indicating cadmium sources at each trophic level exposed to dissolved and dietary cadmium.

Figure 11 suggests that the predator was fed herbivore prey, which also was fed with cadmium-rich food, the algae phytoplankton, so fish and zooplankton were exposed to cadmium in both food and water. Hence, dietary exposure may occur, because the food algae and zooplankton can adsorb and take up dissolved cadmium from the exposure solution before being ingested by the zooplankton and fish.

The development and implementation of effective remedial measures depend on our ability to predict the fate and effects of metals in these systems (Hare and Tessier, 1996). So, cadmium uptake through dietary intake is an important source and the toxicity of dietary cadmium is significantly important for aquatic organisms (Sofyan et al., 2007).

4.1. Phytoplankton Exposed to Cadmium

Cadmium has been ranked as one of the major potential metal hazards, which can produce biological effects in aquatic environments and can increase cell volume, lipid relative volume, and vacuole relative volume in algae. These biological effects may result in structural changes in planktonic communities. In particular, they reduce the richness of micro-algal species, and micro-algal production and alter the micro-algal community structure. The toxic effect of cadmium on microalgae is relevant since these organisms constitute the base of the marine food chain.

Phytoplankton, such as algae, exhibit a net negative charge resulting in an affinity for positively charged species, such as toxic cadmium cations, which will readily adsorb to algal cell surfaces (Taylor et al., 1998) and accumulate significant concentrations of cadmium throughout the food chain (Kremling et al., 1978), posing serious hazards to microalgae. Moreover trace metals such as cadmium typically have a potential affinity for sulfur and nitrogen, and proteins are made up of amino acids, many of which contain sulfur and/or nitrogen, so there is no shortage of potential binding sites for trace metals within algal cells. Such affinities make cadmium potentially toxic, binding to proteins or other molecules and preventing them from functioning in their normal metabolic role (Rainbow, 2002).

4.1.1. Toxic Values of Microalgae Exposure to Cadmium

Table 3 gives different median effective concentrations (EC50, the cadmium concentration that reduces the population growth to 50% of the control) and the rank of various algal species exposed to cadmium, the values of which are highly dependent on and/or are influenced by factors such as strains of algal species, continuous or batch test, composition of the cultures medium, exposure time, and other experimental conditions.

Toxicity values are available for six species of saltwater diatoms, one species of dinoflagellate and green alga, and two species of macroalgae (Table 3). Concentrations causing fifty percent reductions in the growth rates of diatoms range from 60 µg/L for Ditylum brightwelli to 22,390 µg/L for Phaeodactylum tricornutum, the most resistant to cadmium. The brown macroalga (kelp) exhibited mid-range sensitivity to cadmium, with an EC50 of 860 µg Cd/L. The most sensitive saltwater plant tested was the red alga, Champia parvula, with significant reductions in the growth of both the tetrasporophyte plant and female plant occurring at 22.8 µg Cd/L.

Toxicity values are available for six species of saltwater diatoms, one species of dinoflagellate and green alga, and two species of macroalgae (Table 3). Concentrations causing fifty percent reductions in the growth rates of diatoms range from 60 µg/L for Ditylum brightwelli to 22,390 µg/L for Phaeodactylum tricornutum, the most resistant to cadmium. The brown macroalga (kelp) exhibited mid-range sensitivity to cadmium, with an EC50 of 860 µg Cd/L. The most sensitive saltwater plant tested was the red alga, Champia parvula, with significant reductions in the growth of both the tetrasporophyte plant and female plant occurring at 22.8 µg Cd/L.

Table 3. Median effective concentrations and rank of various algal species exposed to cadmium.

Strains of Algal Species		EC50 (µg Cd/L)	Effect	Rank of Sensitivity
Kelp, *Laminana saccharina*		860	8-day growth rate	7
Green alga, **Chlorella pyrenoidosa**		**81.16**	**96-hour growth rate**	**3**
Dinoflagellate, *Prorocentrum micans*		60	30-day growth rate	2
	Phaeodactylum tricornutum	22390	growth	9
	Tetraselmis suecia	7900	6-day growth	8
Diatom	*Asterionella japonica*	224.8	72-hour growth rate	6
	Skeletonema costatum	175	96-hour growth rate	5
	Thalassiosira pseudonana	160	96-hour growth rate	4
	Ditylum brightwellii	60	5-day growth	2
		22.8	Reduced female growth or stopped sexual reproduction	1
Red alga, *Champia parvula*		24.9	Reduced tetrasporophyte growth	*Note*: only data from sensitive stages were used
		77	NOEC sexual reproduction	
		189	Reduced tetrasporangia production	

References: from USEPA, 2001; UNEP, 1985; Nassiri et al. 1997; Mo´nica et al. 2002.

The second sensitive alga was the dinoflagellate Prorocentrum micans, the growth rate of which was inhibited by 1.2 µg Cd/L with resulting cell numbers in the cultures being less than one tenth of the control values. No effect was found at 0.4 µg Cd/L. However other researchers found that the growth rate of this species was only slightly affected at 5 µg Cd/L and then only after 22 days exposure, 50% reduction occurred at 60 µg Cd/L with 30 days exposure. Concentrations greater than 10 µg Cd/L were found to increase the vacuolation and number of lysosomes in this species. Researchers also recorded a reduced growth rate of Isochrysis galbana when exposed to 1 µg Cd/L for 10 days and found that 10 µg Cd/L temporarily reduced the growth rate of Scrippsiella faeroense, but there was no such effect at 2.0 µg Cd/L (UNEP, 1985 and literature cited by this report).

The third sensitive alga was the green alga Chlorella pyrenoidosa, the 96-h EC50 and its 95% CIs (confidence intervals) of which were determined to be 81.16 (71.87–95.12) µg Cd/L for biomass followed ASTM (American Society for Testing and Materials) guidelines without EDTA (ethylenediaminete traacetic acid) addition. These values were similar to the values reported by some other researchers (Lin et al., 2007), although their experimental conditions such as temperature and testing periods were slightly different from these tests.

The reason for the different toxic values above is probably that the exclusion mechanisms and detoxification processes of different strains exposed to cadmium are different (Wang et al., 2009a). Indeed, microalgae can develop a tolerance toward metallic pollutants by the development or the activation of exclusion processes and/or internal sequestration of the pollutants. The exclusion may occur by sequestration and excretion of pollutants in the cell wall or may be a result of the cellular release of organic material that can chelate pollutants in solution. Internal sequestration is generally the result of the production of specific soluble ligands such as metal-binding polypeptides synthesized by algal species from several classes or may occur by the compartmentalization of pollutants within lysosomal/vacuolar systems or precipitation in the form of granules.

4.1.2. Algal Cadmium Burden

Phytoplankton can accumulate significant concentrations of cadmium, although decomposing cells rapidly release cadmium into the water (Kremling et al., 1978). The concentrations of cadmium are usually lower in surface waters than in deep ocean waters, probably because of cadmium uptake into certain species of phytoplankton in surface waters. It is assumed that the metal is loosely bound to the cell's surface, and this process appears to be important in the biogeochemical cycling of cadmium in the marine system.

Single-celled green algae are capable of concentrating cadmium in the cells by both adsorption and active uptake, and the total and intracellular cadmium accumulation increased with aqueous cadmium cations [Cd2+]aq, but the growth rate decreased with the increase of [Cd2+]aq, presumably due to cadmium toxicity, which may additionally result in an increase in the cellular cadmium concentration. The amount accumulated is also directly proportional to the concentration of cadmium present initially and is dependent upon the pH of the medium. Adsorbed cadmium is proportional to [Cd2+]aq over almost three orders of magnitude at constant pH, suggesting a roughly 1:1 stoichiometric interaction between [Cd2+]aq and the surface adsorption sites.

Green alga such as saltwater Chlorella pyrenoidosa accumulated cadmium in a dose-dependent manner (Figure 12), with algal cadmium burdens increasing from less than the detection limit in control treatments to 73.86×10-16 g Cd/cell in algae exposed to 70.29 µg Cd/L, which was related to the adsorption and uptake of cadmium by the growing algae (Wang et al., 2009b).

Metal uptake by living algal cells is said to arise from two sequential processes: an initial rapid passive uptake due to surface binding of ions to the cell walls and a subsequent relatively slow active membrane transport of the metal ions through the wall into the cytoplasm of the cell (Khoshmanesh et al., 1996). Figure 12 shows that in the ranges of 12.9 to 47.1 µg Cd/L, there was a rapid increase in the mass of cadmium adsorbed to the cell as the dissolved cadmium concentration rises, whereas at dissolved cadmium concentrations from 47.1 to 70.3 µg Cd/L, the cadmium uptake in the algae becomes essentially constant and reaches an apparent "plateau". Maybe cellular binding saturation has been reached (Wang and Wang, 2008) or the algal cells have become saturated by adsorbed cadmium. Saturation levels corresponding to 79% surface coverage have been shown for C. pyrenoidosa through the adsorption isotherms (Khoshmanesh et al., 1996). However, cell density was inversely correlated with cadmium exposure, decreasing from 670×104 to 38×104 cells/ml with enhanced cadmium concentrations (Figure 12), so growth rate decreased as cadmium concentration increased in the medium and the inhibition was proportional to cadmium concentration (Wang et al., 2009b).

The ability of Chlorella pyrenoidosa to accumulate large concentrations of cadmium before showing adverse effects may be related to the presence of cadmium-sequestering agents within the cell. Moreover, algae are not the only suspended solid-phase component found in natural waters. The suspended solids in the water are a mixed collection of different materials, including resuspended sediment material, colloids, fecal material, and organic decomposition materials, which can easily adsorb metals and could constitute a source of contamination and toxicity for filter feeders, and could even pose a hazard to the food chain of all aquatic ecosystems.

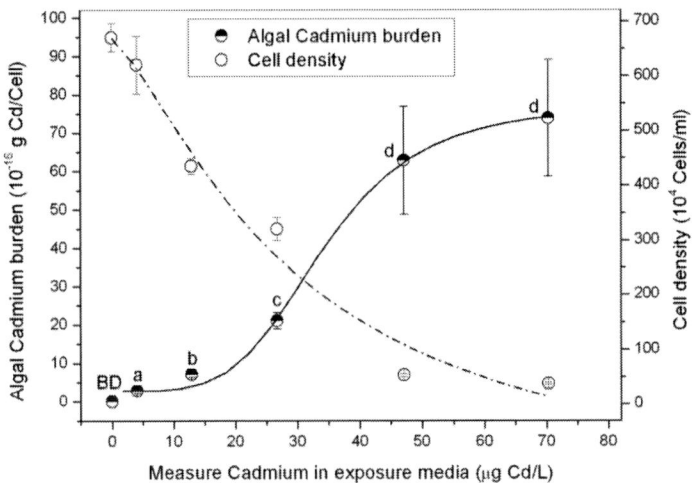

Figure 12. Effects of cell density and algal cadmium burdens of *Chlorella pyrenoidosa* for 96-h exposure to a control and five cadmium concentrations. Vertical error bars represent standard deviation of mean ($n=3$) of algal cadmium burdens and cell density. Means with a different letter are significantly different from one another (two-sided Student's *t* test, $p<0.05$). BD = below the limit of detection.

4.1.3. Cadmium Removed by Unicellular Microalgae

Algae have the ability to concentrate cadmium from aqueous solutions, which has the potential for bioremediation of cadmium-polluted seawater at a lower cost than wastewater treatment processes such as ion exchange, electrochemical treatment or evaporation. Both living and dead algal cells are also known to accumulate metals as "biosorbants", but cadmium was more quickly adsorbed by dried cells than that by living cells.

Cells of the *Chlorella* species are reported to be spherical or ellipsoidal with diameters in the range 2 to 10 µm. Other cells are also spherical elongate or egg-shaped and/or the same size. The mean specific areas were calculated on the assumption that the algal cells are spherical in shape. If we assume that the hydrated ions are essentially spherical, their apparent cross-sectional area can be estimated. The hydrated radius of the cadmium ion has been calculated to be 4.37×10^{-8} cm. Therefore the cross-sectional area of a hydrated ion will be about 6×10^{-5} cm^2, and hence the surface coverage of the ions on the cell at saturation may be estimated from:

Surface coverage (%) = $100 C_1 A N_{AV}/S$,

Where A is the cross-sectional area of a hydrated ion (cm^2 per ion), C_1 is the metal adsorbed on the cell surface, N_{AV} is Avogardro's number and S is the specific surface area (Khoshmanesh et al., 1996).

The uptake cadmium by algae can be calculated through adsorption isotherms, for example, the maximum cadmium adsorption (q_{max}) of marine green alga *Chlorella* sp. NKG16014 was estimated to be 37.0 mg Cd (g dry cells)$^{-1}$ using the Langmuir sorption isotherms (Tadashi et al., 1999).

4.1.4. Mechanism of Cadmium to Microalgae

Cadmium uptake is described in terms of two distinct steps that are believed to occur when cadmium cations are taken from solution by living algal cells: an initial rapid uptake of [Cd^{2+}] due to attachment to the cell wall, followed by a relatively slow uptake due to membrane transport of the ion through the cell wall into the cytoplasm (Khoshmanesh et al., 1996). The first step in cadmium uptake by biota is the adsorption of aqueous ions or complexes on external layers of the algal cell wall, so these "primary" adsorption processes are crucial for modeling the impact on cadmium transport in natural settings and the ecotoxicological consequences.

Similar to other metals, cadmium can interact with both high-affinity (phosphoryl, sulfhydryl) and low-affinity (carboxyl) sites on cell surfaces, depending on its concentration in solution. Complexation with carboxylate surface groups is the primary process responsible for cadmium binding on algal cell walls, so carboxyl and phosphoryl sites play dominant roles in cadmium bio-sorption by marine algae. Furthermore, some detoxification mechanisms via glutathione and phytochelatin (PC*n*) production triggered by cadmium incorporation in the cells appear to play a role both in metal storage and in detoxification since low-level PC*n* concentrations are produced by many species of phytoplankton even at inorganic cadmium contents far below those that impede the growth (Pokrovsky et al., 2008).

The diatom cell wall structure can be viewed as a layer of amorphous silica (frustule) attached to a protein template from the interior of the cell and covered by a polysaccharide layer bearing negatively charged >R-COO moieties according to macroscopic and spectroscopic measurements. Therefore, the metal speciation at the outermost diatom cell walls resulting from short-term adsorption is likely to be controlled by >R-COO-Cd^+ surface complexes, and solely carboxylate moieties were found to be sufficient to provide an adequate fit to the sorption data (Ge'labert et al., 2007). So the order of cadmium toxicity to the diatoms' cell was *S. costatum*>*A. japonica*>*T. suecia*>*P. tricornutum* (EC50, Table 3), in agreement with the decrease of carboxylate group concentration in the surface layers.

Subcellular partitioning of metals may provide a mechanistic approach to investigate metal toxicity and tolerance. The cellular cadmium can be separated into soluble and insoluble fractions, or into the five biologically relevant fractions (namely MRG, cellular debris, organelle, HDP, and HSP), which also were considered to two subcellular compartments comprised of these fractions (MSF and BDM). The presumed metal sensitive fraction was defined as organelles + HDP, and the biological detoxified metal was defined as HSP + MRG. Soluble cadmium was defined as HSP + HDP, and insoluble cadmium was MRG + cellular debris + organelles (Wang and Wang, 2008). For diatoms, HSP was the largest subcellular pool for cadmium, followed by organelles, cellular debris, and MRG, with the least concentration in HDP. Thus, metal detoxification was presumably the major pool for cellular cadmium in diatoms. Metal detoxification can take place by binding to inducible metal-binding proteins or through the precipitation of metals into insoluble forms. Such internal storage and detoxification of metals by MTs and MRG may increase metal tolerance. Further, the 'spillover' of cadmium from these detoxified fractions to other subcellular fractions, i.e., redistribution of cadmium to the sensitive sites, may cause deleterious effects on the photosynthetic PS II system and the growth of the diatoms (Wang and Rainbow, 2006).

Toxicological end points exhibited by diatoms would be more closely related to partitioning of cadmium to organelles and enzymes (the metal-sensitive fractions) than to surface-absorbed or total cellular cadmium (Wallace et al., 2003). Moreover, the correlation between the growth rate inhibition and MSF or organelles was the most significant, both of which strongly demonstrated that cadmium toxicity in diatoms was best predicted by its distribution in MSF or organelles. The reason maybe mainly that the organelles are the principal component in MSF (>87% in most treatments), but it was difficult to determine if cadmium in HDP is a good predictor of cadmium toxicity given its small quantity, because the HDP only made up a small fraction of MSF.

In the diatom, HSP was the most important subcellular fraction for cadmium accumulation and most cadmium was distributed in the insoluble fraction (a combination of metal-rich granules, cellular debris, and organelles). Toxicity differences were the smallest among the different nutrient conditioned cells when the cadmium concentration in the soluble fraction (a combination of HDP and HSP) was used, suggesting that intracellular soluble cadmium may be the best predictor of cadmium toxicity under different nutrient conditions (Miao and Wang, 2006). For example, there was similarly a strong correlation between the cellular accumulation (surface-adsorbed, total cellular and intracellular cadmium) and the growth rate inhibition, and the differences of EC50 based on these parameters were smaller than those of $[Cd^{2+}]$. Furthermore, among the different types of cadmium concentrations, the soluble cadmium was the best predictor of cadmium inhibition of μ and Φ_M for the diatom *T. weissflogii* under different nutrient conditions (Wang and Wang, 2008; Miao and Wang, 2006).

4.2. DIETARY TOXICITY

Cadmium uptake in the aquatic food chain can be considered as occurring through three main vectors:

(i) it can be taken up directly from water column or from interstitial water in sediment for plants and microorganisms;
(ii) sediment-ingesting organisms have a geochemical source of cadmium as well as water source for heterotrophs;
(iii) food items provided by other biota represent a potential source of cadmium as does the water column and diet is an important route for cadmium accumulation in aquatic organisms. Food choice influences body loading, which poses a serious hazard to other consumers.

Previous studies of cadmium toxicity were conducted based on dissolved metal concentrations, with the assumption that toxicity was caused by waterborne metal only. In recent years forever, more and more researchers have considered that cadmium uptake through dietary intake is also an important source for aquatic organisms. The reproductions of marine copepods (*Acartia tonsa* and *Acartia hudsonica*) decreased when fed with the diatom *Thallasiosira pseudonana* contaminated with cadmium (Hook and Fisher, 2002). Toxic effects of reproduction were observed when freshwater cladoceran *Ceriodaphnia dubia* expose to dietary cadmium (Sofyan et al., 2007). Conversely, reproduction and growth increased when the freshwater cladoceran *Daphnia magna* was fed with the cadmium-contaminated algae *Chlamydomonas reinhardtii* and *Pseudokirchneriella subcapit* (Garcia et al., 2004).

Figure 13 shows that dietary cadmium negatively affects net reproductive rate per brood of *M. monogolica*, which is a keystone species in estuarine food webs of China. The net reproductive rate considered two reproductive traits of the brood size and the fraction of parent females producing broods, the dramatic reductions of which from broods 2 to 4 occurred with increasing dietary cadmium and prolonged exposure. Only one female produced neonates or all female were dead in the fourth brood, which demonstrated that the net reproductive rate was severely affected in algal cadmium burdens greater than or equal to 21.06×10^{-16} g Cd/cell (Wang et al., 2009b).

However, when examining the effects of dietary cadmium, no mortality was observed in short term exposures, which were significantly different from those of waterborne cadmium.

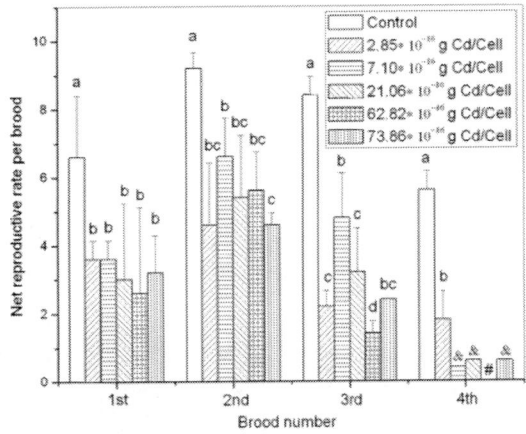

Figure 13. Net reproductive rate per brood of *Moina monogolica* in 21-d dietary cadmium experiments after feeding with *Chlorella pyrenoidosa* exposed for 96 h to a control and five cadmium concentrations. Vertical error bars represent the standard deviations among five replicates. Means with the same letter are not significantly different (Student–Newman–Keuls multiple-range tests, $p<0.05$). The "&" means that only one female produced neonates, so comparisons could not be made because of no standard deviations. The "#" means that all exposed organisms died in the fourth brood before day 21.

The presence of algae in the culture medium could change the speciation and bioavailability of cadmium because of the possible release of algal exudates, which make it difficult to separate effects of dietary and aqueous exposure, particularly for zooplankton, which feed on small particles that rapidly exchange contaminants with water. Moreover, cadmium transfer in aquatic planktonic food chains is controlled by several important physiological and geochemical parameters, including cadmium assimilation and elimination. Fortunately through specific experimental procedures, low cadmium desorption was observed with values below the detection limit, demonstrating that algae were the main cadmium source during dietary exposure.

4.3. MECHANISMS OF DIETARY CADMIUM

For aquatic organisms, it is generally assumed that metal accumulated from water exposure is more likely to be deposited in the respiratory organs (such as gills), whereas metal from particulate-bound form or via the diet following ingestion and digestion is assimilated and mainly deposited in internal tissues.

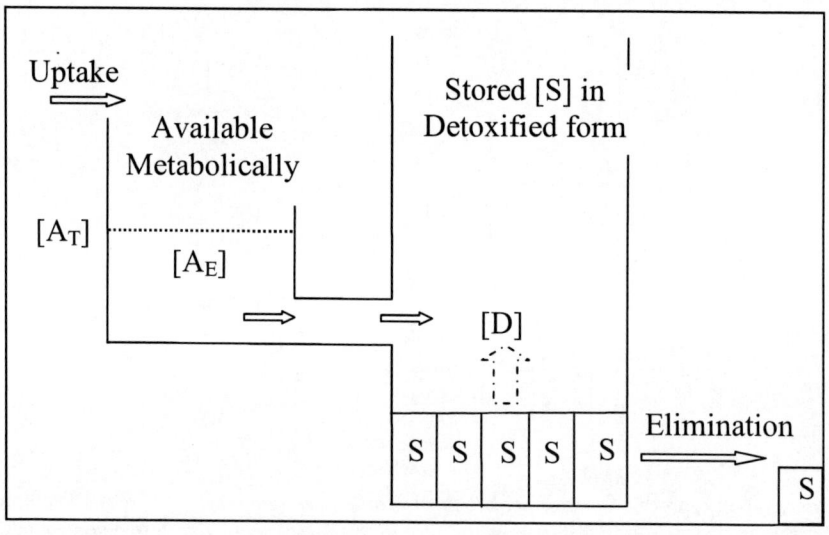

Figure 14. Cadmium accumulation patterns of aquatic organisms from dietary sources showing net accumulation with some excretion of cadmium accumulated in a detoxified form, other details as for Figure 10.

Tolerance or resistance to metal toxicity is based on controlling metal cellular speciation. Sequestration by cellular ligands such as MTs, lysosomes, and mineralized organically based concretions appears to be one of the most commonly adopted strategies by invertebrates (Figure 14).

Because of this differential distribution, dietary metal can exert toxicity via mechanisms different from those of dissolved metal, and, finally, metal distribution and detoxification mechanisms induced by dietary contamination could be different from the contamination of waterborne exposure (Fraysse et al., 2006).

Identification of subcellular metal distribution in aquatic animals and quantification of the subcellular fates of metals in relation to metal toxicity or metal transfer to a higher trophic level can be achieved through different biochemical fractionation techniques (Geffard et al., 2008).

After partitioning, different detoxification processes including detoxification proteins (MTs) and target molecules such as enzymes could be separated into different sub-cellular fractions as insoluble and soluble fractions, the latter often involving binding to physiologically important proteins such as enzymes, or MTs, or the partitioning of cadmium to subcellular compartment containing trophically available metals (TAM) (i.e., HSP, HDP, organelles, 'insoluble' components [e.g., exoskeleton and metal-rich granules] and cellular debris) of specific target tissues or cells (Seeebaugh et al., 2006), especially accumulation in which is most probably related to toxic effects on reproduction (McCarty and Mackay, 1993).

Cadmium distributions into soluble (cytosolic) and insoluble fractions were observed in aquatic organisms of exposure to waterborne and dietary routes, showing their similar potential toxicity. The soluble fraction is made up of MT-like proteins, and also metals binding target molecules such as enzymes, which also seem to have an important role in the detoxification and toxicology of metal. The metal distribution can be related to metal toxicity or trophic transfer. Dietary cadmium was mainly accumulated in the soluble fraction (from 75 to 85% of total accumulated cadmium), which induces deleterious effects on zooplankton reproduction (Figure 14).

Chapter 5

ENVIRONMENTAL SAFETY OF CADMIUM

Sensitivity differences within these "target" groups are often described by statistical distributions using the species sensitivity distributions (SSDs) concept (Figure 15). In this figure, acute cadmium saltwater toxicity data were available for 61 species. Toxicity values ranged from 41.3 µg/L for the Mysid to 135 mg/L for an oligochaete worm. Copepods appeared to be the most sensitive zooplankton and Striped bass was the fish family most sensitive to acute saltwater cadmium exposures.

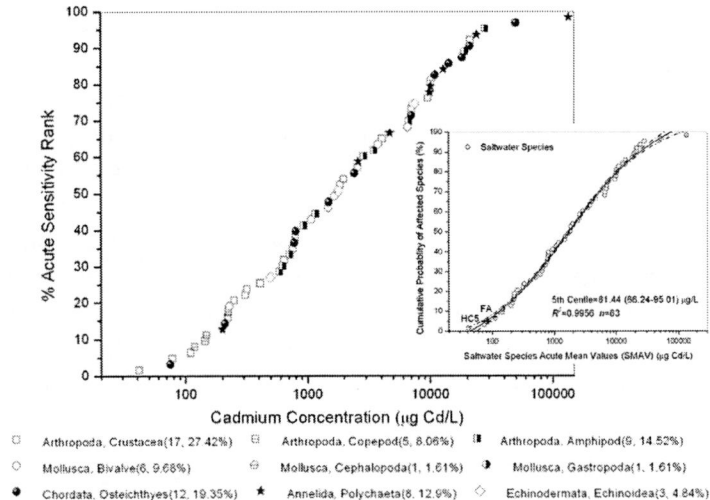

Figure 15. Species Sensitivity Distributions (SSDs)* and Logistic Regression of Various Trophic Saltwater Organisms for Acute Cadmium Toxicity Values.

---

* The species percentile rank in the cumulative frequency distributions for all species toxicity data (i.e., LC50s or SMAVs) and trophic groups were determined by ranking the toxicity value by

The primary toxicity benchmark used for this risk assessment was the 5th percentile of species sensitivity from acute exposures (protection the 95% species). The implied assumption of this approach is that protecting a large percentage of species assemblage will also preserve ecosystem structure and function when using this benchmark. The acute 5th percentile value was selected for the following reasons: (1) based on laboratory experimental data, dissolved and bioavailable cadmium are in the water column of the aquatic environment for only short periods of time that are more closely related to acute exposures that chronic exposures; (2) exposure duration data presented in the "Exposure Characterization" section showed that spike concentrations of cadmium are short-lived (hour to days) in the environment (e.g., cadmium rapidly complexes with natural organic particulates); and (3) toxicity data are much more numerous and represent a wider range of trophic groups for acute studies than chronic studies (Schuler et al., 2008).

The saltwater 5th percentile values (concentration protecting the 95% species) for all species derived from the saltwater acute cadmium toxicity database in figure 15 was 81.44 µg Cd/L, which is below the SMAVs (Species Mean Acute Values) for the mysid, *Mysidopsis bigelowi* (110 µg Cd/L), copepod (adult), *Acartia tonsa* (118.7 µg Cd/L), and cladoceran Moina monogolica (>1.0 mg Cd/L), but is greater than the SMAV for the mysid, *Americamysis bahia* (41.3 µg Cd/L), and aproximately the SMAV of the American lobster, *Homarus americanus* (78 µg Cd/L). Although the saltwater acute cadmium database did not include any macrophyte data and the phytoplankton data were also limited, the most sensitive plant species is more resistant than the chronically most sensitive animal species tested after compared. Therefore, cadmium Water Quality Criteria (WQC) that protects saltwater animals should also protect saltwater plants.

These toxic values of cadmium to various saltwater organisms through waterborne and dietary sources may have important relevance to environmental scenarios in natural aquatic systems as well as consequences on cadmium WQC.

Figures 15 illustrates that while most acute effects occur at cadmium exposures of µg to mg/L, for chronic effects the concentrations of cadmium that

---

concentration. To obtain graphic distributions for each species, the percentile ranking was calculated using the following formula: $Rank = 100 \times \dfrac{n}{N+1}$, where n is the species rank and N is the total number of species observations (Schuler et al., 2008). These percentiles were plotted against log-transformed concentration and logistic regression was fitted to define each distribution. The distributions were plotted graphically using OriginPro 7.5. The 5th centile of the SSDs for cadmium from the acute salt-water data are the "toxicity benchmarks" selected as they are widely accepted and used to characterize risk (Wang, 2007).

evoke the various end points are as low as the μg to ng/L although chronic cadmium toxicity data for saltwater species were few. For example, chronic effects through waterborne routes usually become apparent at concentrations greater than 4.1 μg Cd/L (Wang et al., 2009a). Similarly, natural zooplankton communities in enclosures of 400 ng Cd/L reduced the abundance of crustacean zooplankton communities, while the abundances of the most sensitive cladoceran species were reduced at 200 ng Cd/L and toxicity to *Salmo gairdneri* at less than 500 ng Cd/L. Even some species have been reported to be affected at cadmium concentrations less than 200 ng Cd/L and usually after prolonged exposure under laboratory conditions, because these authors noted sublethal adverse effects on zooplankton at 200 ng Cd/L from related experiments and suspected that fish, especially long-lived species, might show effects in the future because they were still accumulating cadmium (Wright and Welbourn, 1994).

When green alga, *Chlorella pyrenoidosa* are exposed to 4.1 μg Cd/L and resulted in algal cadmium burden of $(2.85\pm0.76)\times10^{-16}$ g Cd /Cell (Figure 12), 59.06% of total neonate production was reduced for *Moina monogolica* exposure to dietary routes (Figure 13).

For risk assessments of the surface saltwater, distributions of actual measured concentrations of cadmium in the Changjiang estuary of China were compared to distributions of toxicity effects data from laboratory studies. These values (Figure 2) were significantly lower than the acute toxicity values (i.e., 5th percentile of species sensitivity and LC50s) and EC50 of algae (Table 3), but greater than the chronic toxicity values (NOEC, LOEC and MATC (maximum allowable toxicant concentrations) from chronic studies) (Wang et al., 2009a). So the potential acute risk to aquatic organisms from cadmium in surface water was low, because acute toxicity generally occurs over the microgram or milligram per liter range. However, chronic risks in salt-water from cadmium exposure in surface water and pore water of sediment is significantly higher than acute risks.

Cadmium chronic effects of waterborne (Wang et al., 2009a) and dietary sources (Figure 13) were compared against Marine Seawater Quality (GB 3097-97) that was issued by China Ministry of Environmental Protection and State Oceanic Administration (CMEP-SOA, 1998) in order to prevent and control marine seawater pollution, protect the marine life and resource, take advantage of marine resource sustainable use, maintain marine ecological equilibrium, and protect human health. GB 3097-97 has three standard criteria for cadmium. The primary seawater standard criteria, which is the most strict as 1.0 μg Cd/L, is applied to protecting habitats for fishery, marine life including natural, rarer, and endangered species. The secondary standard criteria as 5.0 μg Cd/L is applied to regulating foods industrial use, aquaculture, human recreation, and sports, and coastal tourism; and the tertiary standard criteria as 10.0 µg Cd/L is for defining harbors and ocean exploration.

Unfortunately, thresholds (NOEC, MATC and LOEC) (Wang et al., 2009a) all lower than the current WQC of secondary standard criteria (i.e., 5 µg Cd/l), which were compared because *M. monogolica* was cultured in high numbers as a food organism for aquaculture in China. So this value of criteria was considered to be high and may not protect *M. monogolica* or other relevant species from the chronic effects of cadmium.

Furthermore in Canada, the existing guideline for cadmium was also "not sufficient to protect soft water systems from the effects of cadmium because they do not prevent long term accumulation of cadmium in sediments and biota". Guidelines or criteria of cadmium are normally in the order of nanograms per litre for the protection of human health (e.g., in drinking water) or in the 100s of nanogram per liter range for aquatic life based on chronic toxicity (e.g., in water quality guidelines) (Wright and Welbourn, 1994).

Chapter 6

CONCLUSION

Cadmium inputs to the estuarine and marine environment as a probable carcinogen are derived from both natural and anthropogenic sources. The cadmium contents in the estuarine, coastal and marine waters varied between 4.4 and 5.0 µg/L, and a maximum of 10.3 µg Cd/L cadmium has been reported in the Changjiang estuary, China.

Both acute and chronic cadmium exposure can have toxic effects on aquatic organisms. Acute cadmium poisoning can cause death, and chronic (long-term, low level) exposure affects the reproduction, the mechanisms of which were the amount of cadmium accumulated and stored in the body increasing with exposure time and external concentrations. Once cadmium levels in their body (Critical Body Residues, CBRs) reach a particular level (threshold values), toxic effects occurs, initially sublethal but eventually lethal.

Waterborne and particle-bound (dietary) cadmium is bioavailable and toxic to *M. monogolica*. The concentrations inducing adverse effects showed that cadmium is present in aquatic environments at concentrations that are sufficient to provoke a biological response in the natural or indigenous biota.

In this chapter, the concentration protecting 95% saltwater aquatic organisms calculated from the SMAVs is 81.44 µg Cd/L. The important significance when using this benchmark and toxic thresholds is that protecting a large percentage of the species assemblage will preserve ecosystem structure and function. Furthermore, we illustrate the need to take the dietary pathway into account in regulatory assessments and to establish effective concentrations with particulate bound metals of WQC.

Chapter 7

PERSPECTIVES

Cadmium probably manifests their toxic effects by eliminating some species and not affecting others in natural aquatic systems, which seems important to determine which species are most likely to be eliminated. Those species are the useful indicators of cadmium effects and the drivers of how ecosystems will change in response to cadmium contamination. Such information is urgently needed with regard to studies of differences among taxa in toxicity, detoxification, and the resulting responses to cadmium hereafter.

Aquatic organisms are exposed to many different cadmium forms or species such as waterborne and particle-associated cadmium in natural aquatic systems, but WQC assumes that cadmium toxicity primarily occurs via the dissolved phase and no WQC are available for particulate cadmium (in mg/kg) currently. So in future, the importance of considering not only water chemistry but also dietary chemistry in devising environmental regulations for cadmium should be emphasized and the cadmium WQC and guidelines are suggested to be reconsidered.

Toxicity and mechanisms of waterborne and dietary cadmium to saltwater aquatic organisms has been addressed in this chapter based on laboratory controlled experiments of single species populations, which maybe have limitations from an ecological or ecotoxicological perspective, because extrapolation from laboratory to field is difficult and risky. More investigations of population, community, and ecosystem level effects and pathways of cadmium are needed, and risk assessments of cadmium should always include bioassay or investigations of *in-situ* or field community condition in future.

Furthermore, cadmium as the one of the most widely used constituent metal in Quantum dots (QD) core metalloid complexes of Nanoparticles, are known to cause acute and chronic toxicities in organisms and pose environmental and human health risks, but information on routes of exposure and the environmental transport and fate of cadmium QD materials is scarce. The potential risks by which to the environment should be seriously considered.

ACKNOWLEDGMENTS

This work was funded by the Knowledge Innovation Program of Chinese Academy of Sciences (CAS) (Series No. KZCX2-YW-422-4 and No. 0714091d20) and National Natural Science Foundation of China (Grant No. 20807034/B070704 to Zaosheng Wang). The authors would like to thank Prof. Dr. Dave Mount (U.S. EPA) for the professional and beneficial suggestions. Thanks are extended to Prof. Dr. Jingzhu Zhao and Prof. Dr. Yongguan Zhu for leading and managing the project of IUE (Institutions of Urban Environment) as well as for scientific assistance.

Reviewed by Prof. Dr. Graeme Batley, Chief Research Scientist, Centre for Environmental Contaminants Research (CECR), Commonwealth Scientific and Industrial Research Organization (CRISO) Land and Water, New Illawarra Road, Lucas Heights, Private Mai Bag 7 Bangor NSW 2234, Australia.

ABBREVIATIONS

ASTM	American Society for Testing and Materials
BCF	Bio-concentration Factor
BDM	Biologically Detoxified Metal
BLM	Biotic Ligand Model
CBRs	Critical Body Residues
DOC	Dissolved Organic Carbon
DYMBAM	Dynamic Multi-pathway Bioaccumulation Model
EDTA	Ethylenediaminetetraacetic acid
GESAMP	Group of Experts on the Scientific Aspects of Marine Pollution
HDP	Heat-Denatured Protein
HSP	Heat-Stable Protein
LOEC	Lowest-Observed-Effect Concentration
MATC	Maximum Allowable Toxicant Concentrations
MRG	Metal Rich Granules
MSF	Metal-Sensitive Fractions
MTs	Metallothioneins
NOEC	No-Observed-Effect Concentration
PCn	Phytochelatin
POC	Particulate Organic Carbon
SMAVs	Species Mean Acute Values
SPM	Sub-cellular Partitioning Model
SSDs	Species Sensitivity Distributions
WQC	Water Quality Criteria

REFERENCES

[1] Campbell PGC. 1995. Interactions between trace metals and aquatic organisms: A critique of the free-ion activity model. In Tessier A, Turner DR, eds, *Metal Speciation and Bioavailability in Aquatic Systems*. John Wiley & Sons, New York, NY, USA, pp 45–102.

[2] Campbell PGC, (2006). Cadmium–a priority pollutant. *Environmental Chemistry, 3*, 387-388.

[3] CMEP-SOA (China Ministry of Environmental Protection, State Oceanic Administration), (1998). The People's Republic of China National Standards GB 3097–97 marine seawater quality. China Ministry of Environmental Protection and State Oceanic Administration, Beijing, China

[4] Culbard EB, Thornton I, Watt J, Wheatley M, Moorcroft S, Thompson M, (1988). Metal contamination in British urban dusts and soils. *Journal of Environmental Quality, 17*, 226-234.

[5] Di Toro DM, Allen HE, Bergman H, Meyer J, Paquin PR, Santore RC, (2001). A biotic ligand model of the acute toxicity of metals. I. Technical basis. *Environmental Toxicology and Chemistry, 20*, 2383–2396.

[6] Di Toro DM, Mahoney JD, Hansen DJ, Scott JK, Hicks MB, Mayro SM, Redinan MS, (1990). Toxicity of cadmium in sediments: the role of acid volatile sulfide. *Environmental Toxicology and Chemistry, 9*, 1487-1502.

[7] Engel DW, Brouwer M, (1987). Metal regulation and molting in the blue crab, *Callinectes sapidus*: metallothionein function in metal metabolism. *Biological Bulletin, 173*, 239-25 1.

[8] Fraysse B, Geffard O, Berthet B, Que´au H, Biagianti-Risbourg S, Geffard A. (2006). Importance of metallothioneins in the cadmium detoxification process in *Daphnia magna*. *Comp Biochem Physiol C, 144*, 286–293.

[9] Garcia GG, Nandini S, Sarma SSS, (2004). Effect of cadmium on the population dynamics of *Moina macrocopa* and *Macrothrix triserialis* (Cladocera). *Bulletin of Environmental Contamination and Toxicology*, *72*, 717-724

[10] Geffard O, Geffard A, Chaumot A, Vollat B, Alvarez C, Tusseau-vuillemin MH, Garric J, (2008). Effects of chronic dietary and waterborne cadmium exposures on the contamination level and reproduction of Daphnia magna. *Environmental Toxicology and Chemistry*, *27*, 1128-1134.

[11] Ge´labert A, Pokrovsky OS, Schott J, Boudou A, Feurtet-Mazel A, (2007). Cadmium and lead interaction with diatom surfaces: a combined thermodynamic and kinetic approach. *Geochim. Cosmochim. Acta*, *71*, 3698–3716.

[12] Hare L, Tessier A, (1996). Predicting animal cadmium concentrations in lakes. *Nature*, *380*, 430–432.

[13] Hook SE, Fisher NS, (2002). Relating the reproductive toxicity of five ingested metals in calanoid copepods with sulfur affinity. *Marine Environmental Research*, *53*, 161–174.

[14] Huebert DB, Shay JM, (1992). The effect of EDTA on cadmium and zinc uptake and toxicity in *Lemna trisulca*. *Archives of Environmental Contamination and Toxicology*, *22*, 313-318.

[15] Khoshmanesh A, Lawson F, Prince IG, (1996). Cadmium uptake by unicellular green microalgae. *The Chemical Engineering Journal*, *62*, 81–88.

[16] Kneer R, Zenk MH, (1992). Phytochelatins protect plant enzymes from heavy metals poisoning. *Phytochemistry*, *8*, 2663–2667.

[17] Kremling K, Piuze J, Brockel VK, Wong CS, (1978). Studies on the pathways and effects of cadmium in marine plankton communities in experimental enclosures. *Marine* Biology, 48, 1-10.

[18] Lee BG, Griscom SB, Lee JS, Choi HJ, Koh CH, Luoma SN, Fisher NS, (2000). Influence of dietary uptake and reactive sulfides on metal bioavailability from aquatic sediments. *Science*, *287*, 282-284.

[19] Lin KC, Lee YL, Chen CY, (2007). Metal toxicity to *Chlorella pyrenoidosa* assessed by a short-term continuous test. *Journal of Hazardous Materials*, *142*, 236–241.

[20] Luoma SN, Rainbow PS, (2005). Why is metal bioaccumulation so variable? Biodynamics as a unifying concept. *Environmental Science and Technology*, *7*, 1921-1931.

[21] Mackay, (1983). Metal organic complexes in sea water. An investigation of naturally-occurring complexes of Cu, Zn, Fe, Mn, Ni, Mg, and Cd using

high performance liquid chromatography with atomic fluorescence detection. *Marine Chemistry, 2,* 169-180.

[22] Miao AJ, Wang WX, (2006). Cadmium toxicity to two marine phytoplankton under different nutrient conditions. *Aquatic Toxicology, 78,* 114–126.

[23] Mo'nica PR, Julio AA, Concepcio'n HL, Enrique TV, (2002) Cadmium removal by living cells of the marine microalgae *Tetraselmis suecica. Bioresource Technology, 84,* 265–270.

[24] Munger C, Hare L, (1997). Relative importance of water and food as cadmium sources to an aquatic insect (*Chaoborus punctipennis*): Implications for predicting Cd bioaccumulation in nature. *Environmental Science and Technology, 31,* 891–895.

[25] Nassiri Y, Mansot JL, We'ry J, Ginsburger-vogel T, Amiard JC, (1997) Ultrastrucal and electron energy loss spectroscopy studies of sequestration mechanisms of Cd and Cu in the marine diatom *Skeletonema costatum. Archives of Environmental Contamination and Toxicology, 33,*147–155.

[26] Nebeker AV, Cairns ST, Oryukka MA, Krawczyk DF, (1986). Survival of *Daphnia magna* and *Hyalella azteca* in cadmium spiked water and sediment. *Environmental Toxicology and Chemistry, 5,* 933-938.

[27] Nriagu JO, Pacyna JM, (1988). Quantitative assment of world wide contamination of air, water and soils by trace metals. *Nature, 333,* 134-139.

[28] Organisation for Economic Co-operation and Development (OECD), (1994). Risk Reduction Monograph No. 5: Cadmium OECD Environment Directorate, Paris, France.

[29] Paquin PR, Gorsuch JW, Apte SC, Bowles KC, Batley GE, Campbell PGC, Delos C, Di Toro DM, Dwyer R L, Galvez F, Gensemer RW, Goss GG, Hogstrand C, Janssen CR, McGeer JC, Naddy RB, Playle RC, Santore RC, Schneider U, Stubblefield W A, Wood C M, and Wu KB. (2002). The biotic ligand model: a historical overview. *Comparative Biochemistry and Physiology,* 133C, 3-36.

[30] Pinot F, Kreps SE, Bachelet M, Hainaut P, Bakonyi M, Polla BS, (2000). Cadmium in the environment: sources, mechanisms of biotoxicity, and biomarkers. *Reviews on Environmental Health, 3,* 299-323.

[31] Pokrovsky OS, Pokrovski GS, Feurtet-mazel A. (2008). A structural study of cadmium interaction with aquatic microorganisms. *Environmental Science and Toxicology, 42,* 5527–5533.

[32] Rainbow PS, (2002). Trace metal concentrations in aquatic invertebrates: why and so what? *Environmental Pollution, 120,* 497–507.

[33] Schuler LJ, Hoang TC, Rand GM (2008). Aquatic risk assessment of copper in freshwater and saltwater ecosystems of south Florida. *Ecotoxicology, 17*, 642-659.

[34] Seebaugh DR, Estephan A, Wallace WG, (2006). Relationship between dietary cadmium absorption by Grass shrimp (*Palaemonetes pugio*) and trophically available cadmium in amphipod (*Gammarus lawrencianus*) prey. *Bulletin of Environmental Contamination and Toxicology, 76*, 16–23.

[35] Sofyan A, Price JD, Birge JW, (2007). Effects of aqueous, dietary and combined exposures of cadmium to *Ceriodaphnia dubia*. *Science of Total Environment, 385*, 108–116.

[36] Tadashi M, Haruko T, Takashi N, Akira Y, (1999). Screening of marine microalgae for bioremediation of cadmium-polluted seawater. *Journal of Biotechnology, 70*, 33-38.

[37] Taylor G, Baird DJ, Soares AMVM, (1998). Surface binding of contaminants by algae: consequences for lethal toxicity and feeding to *Daphnia magna* Straus. *Environmental Toxicology and Chemistry, 17*, 412–419.

[38] Thornton I, (1992). Sources and pathways of cadmium in the environment. *International Agency for Research in Cancer, 2*, 149-162.

[39] UNEP (United Nations Environment Program), (1985). GESAMP (the Joint Group of experts on the Scientific Aspects of Marine Pollution): Cadmium, lead and tin in the environment.

[40] USEPA (US Environmental Protection Agency), (2001). 2001 update of ambient water quality criterion for cadmium. EPA 822-R-01-001. US Environmental Protection Agency, Office of Water Science and Technology, Washington, DC

[41] Vijver MG, Van Gestel CAM, Lanno RP, Van Straalen NM, Peijnenburg WJGM, (2004). Internal metal sequestration and its ecotoxicological relevance: a review. *Environmental Science and Toxicology, 18*, 4705-4712.

[42] Wallace WG, Lee BG, Luoma SN, (2003). Subcellular compartment alization of Cd and Zn in two bivalves. I. Significance of metal-sensitive fractions (MSF) and biologically detoxified metal (BDM). *Marine Ecological Progress Series, 249*, 183–197.

[43] Wang MJ, Wang WX, (2008). Cadmium toxicity in a marine diatom as predicted by the cellular metal sensitive fraction. *Environmental Science and Technology, 42*, 940–946.

[44] Wang WX, Fisher NS, (1999). Assimilation efficiencies of chemical contaminants in aquatic invertebrates: A synthesis. *Environmental Toxicology and Chemistry, 18*, 2034–2045.

[45] Wang WX, Rainbow PS, (2006). Subcellular partitioning and the prediction of cadmium toxicity to aquatic organisms. *Environmental Chemistry*, *3*, 395–399.

[46] Wang ZS, (2007). Study of copper toxicity to a saltwater cladoceram *Moina monogolica* Daday and its methods of toxicity evaluation. Philosopher Doctor Dissertation, Shanghai Jiao Tong University, Shanghai, China.

[47] Wang ZS, Kong HN, Wu DY, (2007). Reproductive toxicity of dietary copper to a saltwater cladoceran *Moina monogolica* Daday. *Environmental Toxicology and Chemistry*, *26*, 126–131.

[48] Wang ZS, Yan CZ, Zhang X, (2009a). Acute and chronic cadmium toxicity to a saltwater cladoceran *Moina Monogolica* Daday and its relative importance. *Ecotoxicology*, *18*, 47-54.

[49] Wang ZS, Yan CZ, Ross VH, (2009b). Effects of dietary cadmium exposure on reproduction of saltwater cladoceran *Moina Monogolica* Daday: implications in water quality criteria. *Environmental Toxicology and Chemistry*, (In press).

[50] Wright DR, Welbourn PM, (1994). Cadmium in the aquatic environment: a review of ecological, physiological, and toxicological effects on biota. *Environmental Reviews*, *2*, 187-214.

INDEX

A

absorption, 62
acid, 3, 13, 26, 37, 57, 59
activation, 37
acute, x, 15, 16, 17, 18, 28, 47, 48, 49, 51, 53, 59
adaptation, 11
adsorption, 29, 38, 40, 41
adsorption isotherms, 38, 40
adult, 48
aerobic, 12
agent, 10
agents, 13, 27, 38
agricultural, 6, 7
air, 4, 5, 6, 61
algae, 34, 38, 40, 43, 49, 62
Algal, 36, 37
alkaline, 5
alkalinity, 11
alloys, 5
ambient air, 6
amino acids, 24, 35
amorphous, 41
anhydrase, 26
animal tissues, 4
animals, 10, 18, 45, 48
antagonistic, 10, 11
anthropogenic, 1, 3, 5, 6, 51
application, 6
aquaculture, 49, 50
aquatic systems, 21, 48, 53
aqueous solution, 39
aqueous solutions, 39
Arthropoda, ix, 16, 18
assimilation, 44
ASTM, 37, 57
Atlantic, 4
atmosphere, 5
atmospheric deposition, 6, 7, 9
attachment, 40
Australia, 55
availability, 10, 11, 12, 26, 30

B

bacteria, 15
batteries, 5
behavior, 4, 5
Beijing, 59
benchmark, 48, 51
benchmarks, 48
bicarbonate, 11
binding, 2, 10, 11, 12, 13, 24, 25, 26, 27, 28, 29, 30, 31, 35, 37, 38, 40, 41, 45, 62
bioaccumulation, 2, 12, 21, 27, 28, 33, 60, 61

bioassay, 53
bioavailability, 3, 10, 13, 18, 43, 60
biological responses, 27
biomarkers, 61
biomass, 37
bioremediation, 39, 62
biota, x, 9, 10, 15, 22, 40, 42, 50, 51, 63
biotic, 2, 11, 26, 27, 59, 61
bivalve, 17, 21, 29
BLM, 11, 27, 57
body fluid, 12
broad spectrum, 15
burning, 1
by-products, 1

C

Ca^{2+}, 12
cadmium, ix, 1, 2, 3, 4, 5, 6, 7, 8, 9, 10, 11, 12, 13, 15, 16, 17, 18, 19, 20, 21, 22, 23, 24, 25, 26, 28, 29, 30, 31, 33, 34, 35, 36, 37, 38, 39, 40, 41, 42, 43, 44, 45, 47, 48, 49, 50, 51, 53, 59, 60, 61, 62, 63
calcium, 3, 11, 27
CAM, 62
Canada, 50
carbon, 3, 10, 11, 13
Carbon, 57
carbonates, 12
carboxyl, 40
carcinogen, 51
CAS, 55
cation, 3, 10
cell, 12, 25, 29, 34, 35, 37, 38, 39, 40, 41, 43
cell surface, 35, 40
channels, 26
chelates, 10
chemical composition, 5
children, 6
China, 7, 8, 9, 17, 18, 43, 49, 50, 51, 55, 59, 63
chloride, 3, 10, 11, 13, 21
Chordata, ix, 17, 18, 19
cladocerans, 20

classes, ix, 37
coal, 1, 6
coastal areas, 9
colloids, 39
combined effect, 12
combustion, 1, 6
communities, 34, 49, 60
community, 34, 53
competition, 11, 25, 27
components, 22, 45
composition, 5, 35
concentration, 1, 2, 3, 4, 6, 8, 13, 20, 21, 23, 24, 25, 27, 28, 30, 35, 38, 40, 41, 42, 48, 51, 57
conceptual model, 27
confidence, 37
consumers, 33, 42
consumption, 5
contaminants, 44, 62
contamination, 6, 9, 39, 45, 53, 59, 60, 61
control, 2, 20, 24, 35, 37, 38, 39, 43, 49
copepods, 16, 19, 42, 60
copper, 24, 28, 30, 62, 63
correlation, 33, 41, 42
correlations, 27
corrosion, 5
crab, ix, 18, 29, 59
Crassostrea gigas, 17, 18, 29
crops, 5, 6
cross-sectional, 39, 40
crust, 3
crustaceans, 21, 28, 29
culture, 43
cumulative frequency, 47
cycling, 1, 38
cysteine, 24
cytoplasm, 38, 40
cytosol, 24, 28, 29
cytosolic, 25, 45

D

database, 48

death, 51
decomposition, 1, 39
density, 38, 39
deposition, 6, 7, 9, 29
deposits, 26
desorption, 44
detection, 38, 39, 44, 61
detoxification, ix, 19, 22, 24, 37, 40, 41, 45, 53, 59
Detoxification, 24
detoxifying, 24
diatoms, 35, 41
diet, 22, 30, 42, 44
dietary, ix, 2, 5, 15, 33, 34, 42, 43, 44, 45, 48, 49, 51, 53, 60, 62, 63
dietary intake, 34, 42
digestion, 44
discharges, 7
dispersion, 4
displacement, 24
distribution, 2, 3, 10, 20, 24, 25, 27, 28, 41, 45, 48
diversity, 9, 28
DNA, 26
dominance, 12
drainage, 6, 7
drinking water, 50
duration, 48
dusts, 4, 6, 59

E

earth, 3
ecological, ix, 27, 49, 53, 63
economic development, 7
ecosystem, 48, 51, 53
ecosystems, ix, 3, 9, 21, 39, 53, 62
effluent, 7
effluents, 5, 6, 9
egg, 39
electron, 4, 61
electronic waste, 5
electroplating, 1, 5

emission, 3
energy, 1, 61
environment, ix, 1, 3, 4, 5, 6, 7, 10, 11, 12, 26, 48, 54, 61, 62, 63
environmental conditions, 21
Environmental Protection Agency, 62
environmental regulations, 53
enzymes, 2, 25, 26, 41, 45, 60
EPA, 55, 62
epithelial transport, 30
epithelium, 12
equilibrium, 27, 49
estuarine, x, 7, 9, 10, 11, 43, 51
Europe, 7
evaporation, 39
exclusion, 37
excretion, 23, 24, 28, 29, 37, 44
exoskeleton, 29, 45
experimental condition, 35, 37
exposure, x, 2, 5, 6, 15, 16, 18, 20, 21, 24, 26, 27, 28, 29, 30, 33, 34, 35, 37, 38, 39, 43, 44, 45, 48, 49, 51, 54, 63
Exposure, 21, 35, 48
extrapolation, 53

F

family, 47
feeding, 5, 17, 43, 62
females, 43
ferritin, 26
fertilizers, 1, 5, 6
filter feeders, 39
fish, ix, 5, 15, 16, 18, 21, 27, 30, 34, 47, 49
fishing, 1
fluid, 12
fluorescence, 61
food, 2, 6, 15, 16, 33, 34, 35, 39, 42, 43, 44, 50, 61
fossil, 6
fractionation, 25, 28, 45
France, 61
freshwater, 7, 10, 11, 12, 18, 21, 42, 62

fuel, 6

G

geochemical, 4, 21, 42, 44
gill, 11, 27, 29
gland, 30
glass, 5
glutathione, 40
granules, ix, 2, 23, 24, 25, 26, 27, 29, 30, 37, 42, 45
grass, 21
ground water, 4
groups, 27, 30, 40, 47, 48
growth, 20, 35, 36, 37, 38, 40, 41, 42, 43
guidelines, 37, 50, 53
gut, 30

H

half-life, 30
hardness, 3, 11, 18
harm, ix, 1, 4
harmful effects, 2, 19
hazards, 1, 2, 5, 34, 35
health, 1, 2, 33, 49, 50, 54
heat, 2, 24, 25, 26, 27
heavy metals, 7, 60
hemoglobin, 26
hemolymph, 29
house dust, 6
HSP, 2, 27, 41, 42, 45, 57
human, ix, 1, 2, 3, 4, 9, 49, 50, 54
humans, 5, 6
humate, 21
humic acid, 10, 12
hydrogen, 12, 27
hydroxide, 11, 13

I

implementation, 34

in vitro, 30
inclusion, 26
inclusion bodies, 26
indicators, 20, 53
indigenous, 18, 51
induction, 25, 30
industrial, 1, 5, 6, 7, 9, 49
industrialization, 7
industrialized countries, 5
industry, 2
inert, 23
ingest, 6, 10
ingestion, 5, 44
inhalation, 5
inhibition, 38, 41, 42
Innovation, 55
inorganic, 10, 11, 12, 21, 29, 40
integrity, 11
interaction, 11, 25, 38, 60, 61
interactions, 12, 24
interstitial, 42
invertebrates, 16, 21, 28, 30, 44, 61, 62
ion channels, 26
ionic, 10, 26
ionization, 4
ionization potentials, 4
ions, 3, 11, 12, 26, 27, 38, 39, 40
iron, 1, 12
isomers, 24, 28
isotherms, 38, 40

J

Japan, 7

K

kelp, 35
kidney, 29, 30

L

laboratory studies, 49
lakes, 7, 60
land, 6, 7
landfill, 9
Langmuir, 40
leaching, 9
life cycle, 20
life-cycle, 20
ligand, 2, 11, 26, 27, 59, 61
ligands, 2, 10, 12, 25, 37, 44
limitations, 53
links, 21
lipid, 34
liquid chromatography, 61
liver, 30
localization, 25, 30
low molecular weight, 24, 26
low-level, 20, 40
lysosomes, 25, 29, 37, 44

M

macroalgae, 35
magnesium, 3, 11, 27
mammals, 5
manganese, 3, 11
marine environment, x, 1, 51
measures, 34
median, 35
metabolic, 22, 23, 24, 26, 35
metabolism, 22, 24, 26, 30, 59
metal ions, 38
metallothioneins (MTs), 23, 24, 57, 59
metals, 1, 2, 7, 11, 12, 15, 21, 22, 23, 24, 25, 26, 27, 33, 34, 35, 39, 40, 41, 45, 51, 59, 60, 61
microalgae, 34, 35, 37, 60, 61, 62
micrograms, 3
microorganisms, 42, 61
microsomes, 25

mineralization, 7
mineralized, 44
minerals, 3, 4
mining, 5, 6, 7
Ministry of Environment, 49, 59
mitochondrial, 25
mobility, 4
modeling, 40
models, 2, 19
moieties, 41
molecular weight, 24, 26, 30
molecules, 2, 23, 26, 35, 45
molting, 59
mortality, 22, 43
MTs, x, 23, 24, 25, 28, 29, 30, 31, 41, 44, 45, 57
mucosa, 30
mucous cells, 30
municipal sewage, 6
muscle, 21, 30

N

Na^+, 12
native species, 18
natural, ix, 1, 3, 4, 5, 10, 20, 38, 40, 48, 49, 51, 53
natural environment, ix
neonate, 20, 49
neonates, 20, 43
neurotransmission, 30
New York, 59
Ni, 60
nickel, 5
nitrogen, 35
non-ferrous metal, 2, 6
nontoxic, 24
normal, 5, 18, 35
normal conditions, 5
norms, x
nuclear, 5, 25
nuclei, 25
nucleus, 24

nutrient, 15, 42, 61

O

observations, 33, 48
oceans, 4, 7
OECD, 6, 61
olfactory, 30
ores, 1, 6
organ, 28, 29
organelle, 41
organelles, 2, 25, 27, 41, 42, 45
organic, 3, 10, 11, 12, 13, 21, 26, 37, 39, 48, 60
organic matter, 12, 21
Organisation for Economic Co-operation and Development, 61
organism, 2, 15, 16, 19, 21, 23, 24, 25, 26, 28, 50
oxide, 4
oxides, 12
oyster, ix, 17, 21
oysters, 5, 29

P

Pacific, 4, 7, 17
paints, 5, 6
Paris, 61
particles, 10, 33, 44
particulate matter, 10, 13, 33
passive, 31, 38
pathways, 12, 15, 53, 60, 62
percentile, 47, 48, 49
Periodic Table, 1
pesticides, 1
pH, 3, 10, 12, 38
phosphors, 5
phosphorus, 6, 29
photosynthetic, 41
phylum, 18
physicochemical, 28

physicochemical properties, 28
physiological, 12, 19, 44, 63
physiology, 28
phytoplankton, 5, 15, 27, 33, 34, 37, 40, 48, 61
pigments, 1, 5
plankton, 60
plants, 5, 6, 24, 42, 48
plasma, 12
plasma membrane, 12
plastics, 5, 6
play, 24, 30, 33, 40
poisoning, 51, 60
pollutant, 37, 59
pollution, 1, 7, 49
polymers, 5
polypeptides, 37
polysaccharide, 41
population, 6, 20, 35, 53, 60
population growth, 20, 35
pore, 12, 49
power, 6
power stations, 6
precipitation, 5, 6, 37, 41
prediction, 63
pressure, 25
pristine, 3
production, 1, 5, 25, 30, 34, 36, 37, 40, 49
program, 9
protection, 5, 48, 50
protein, 2, 25, 26, 27, 41
proteins, 2, 12, 23, 24, 25, 26, 27, 29, 30, 35, 41, 45
Proteins, 24
pulse, 30
PVC, 5
PVC polymers, 5

R

radius, 40
rain, 7

Index

range, 3, 4, 5, 6, 7, 10, 15, 16, 17, 18, 19, 21, 35, 39, 43, 48, 49, 50
recreation, 49
redistribution, 27, 41
redox, 13
refining, 1, 5, 6
regression, 48
regulation, 23, 24, 27, 28, 59
regulators, 24
relationship, 11, 22
relevance, 48, 62
reproduction, 20, 36, 42, 45, 51, 60, 63
residues, ix, 21, 27
resistance, 18, 44
resources, 1
respiratory, 44
risk assessment, ix, 48, 49, 53, 62
risks, 22, 49, 54
rivers, 7
rods, 5
runoff, 7, 9
rural areas, 6

S

saline, 3, 11
salinity, 3, 10, 18, 20
salt, 21, 48, 49
saltwater, ix, 9, 13, 16, 17, 18, 20, 21, 35, 38, 47, 48, 49, 51, 53, 62, 63
sampling, 7, 8, 9
saturation, 38, 40
sea urchin, ix
seawater, 3, 8, 10, 39, 49, 59, 62
sediment, 4, 9, 13, 33, 39, 42, 49, 61
sediments, 3, 4, 7, 9, 12, 15, 21, 50, 59, 60
seed, 5
seeps, 6
selenium, 3, 12
sensitivity, ix, 2, 18, 22, 25, 30, 35, 47, 48, 49
sequestration proteins, 26
sewage, 6, 7
sexual reproduction, 36

Shanghai, 7, 63
shape, 39
short period, 48
shortage, 35
short-term, x, 41, 60
shrimp, ix, 16, 18, 21, 62
silica, 41
sites, 2, 7, 8, 9, 10, 11, 24, 25, 26, 27, 28, 29, 30, 31, 35, 38, 40, 41
sludge, 6
smelters, 6
smelting, 1
soils, 3, 4, 5, 6, 7, 59, 61
solid phase, 12
solid waste, 5
sorption, 40, 41
speciation, 10, 11, 18, 21, 27, 41, 43, 44
species, ix, 10, 12, 16, 17, 18, 19, 20, 21, 22, 26, 27, 28, 30, 31, 34, 35, 36, 37, 39, 40, 43, 47, 48, 49, 50, 51, 53
specific surface, 40
spectroscopy, 61
spectrum, 15
sports, 49
stabilizers, 5
stages, 36
standard deviation, 39, 43, 59
steel, 1
stoichiometry, 24
storage, 22, 23, 26, 28, 40, 41
strains, 35, 37
strategies, 28, 44
streams, 7
strength, 1, 24
stress, 18, 20
structural changes, 34
Subcellular, 2, 27, 41, 45, 62, 63
substances, 1
sulfate, 11, 13, 21
sulfur, 12, 29, 35, 60
superoxide dismutase, 26
supply, 20
surface layer, 8, 41

surface water, 9, 37, 49
survival, 20
synthesis, 62

T

targets, 25
taxa, 28, 53
taxonomic, 16
temperature, 37
thermodynamic, 60
thin film, 5
threat, 2
threshold, x, 23, 24, 25, 51
tin, 62
tissue, 21, 24, 25, 28, 29, 30
tobacco smoke, 5, 6
tolerance, ix, 18, 25, 37, 41
total organic carbon, 13
tourism, 50
toxic, ix, 2, 3, 10, 12, 15, 18, 19, 20, 22, 23, 24, 25, 26, 27, 28, 29, 34, 37, 45, 48, 51, 53
toxic effect, ix, 2, 15, 23, 24, 27, 34, 45, 51, 53
toxic metals, 26
toxicities, 13, 53
toxicity, ix, 2, 3, 10, 11, 13, 15, 16, 17, 18, 19, 20, 22, 24, 25, 26, 27, 28, 33, 34, 39, 41, 42, 44, 45, 47, 48, 49, 50, 53, 59, 60, 61, 62, 63
toxicological, 2, 63
toxicology, 22, 45
transfer, 7, 25, 30, 44, 45
transformation, 3
transition, 24, 25
transition metal, 24
transport, x, 3, 7, 12, 26, 30, 38, 40, 54
transportation, 6
trout, 30

U

UNEP, 7, 19, 36, 37, 62
United Kingdom, 7
United Nations Environment Program, 62
urban areas, 6
urbanization, 6
urbanized, 7
USEPA, 3, 17, 19, 20, 21, 36, 62

V

vacuole, 34
values, ix, 5, 17, 18, 35, 37, 44, 47, 48, 49, 51
variability, 20
variables, 20
volcanic emissions, 5

W

waste disposal, 7
wastes, 6
wastewater treatment, 5, 39
water, ix, 1, 3, 4, 5, 6, 7, 9, 10, 11, 15, 16, 21, 29, 30, 33, 34, 37, 38, 42, 44, 48, 49, 50, 53, 60, 61, 62, 63
water quality, 50, 62, 63
weathering, 5
wood, 1
worm, ix, 17, 18, 47

Y

yield, 20

Z

zinc, 1, 3, 4, 5, 6, 12, 24, 28, 30, 60
zooplankton, 11, 15, 17, 20, 34, 44, 45, 47, 49